LOS ANGELES RIVER 2012
STATE OF THE WATERSHED REPORT

Kristy Morris, Ph.D., *Council for Watershed Health*
Scott Johnson, M.S., *Aquatic Bioassay and Consulting Laboratories Inc.*
Nancy Steele, D.Env., *Council for Watershed Health*

Prepared for:
City of Los Angeles, LA Sanitation
City of Burbank Public Works Department

THE CITY OF LOS ANGELES, LA SANITATION

VISION: An organization that sets the benchmark for outstanding service and responds to the challenges of tomorrow.

MISSION: To protect public health and the environment.

"Working hard every day for a sustainable LA"

THE CITY OF BURBANK PUBLIC WORKS DEPARTMENT

The key objectives of the Public Works Department are to provide for the efficient operation of public works systems and programs such as wastewater treatment, sewer maintenance, street design and maintenance, street sweeping, solid waste collection, recycling and landfill disposal, public building maintenance, equipment maintenance, traffic and parking management, and graffiti removal; while protecting the environment and responding to the changing needs of the citizens.

Acknowledgements

PRINCIPAL AUTHORS
Kristy Morris, Ph.D., *Council for Watershed Health*
Scott Johnson, M.S., *Aquatic Bioassay and Consulting Laboratories Inc.*
Nancy Steele, D.Env., *Council for Watershed Health*

Thank you to the following Los Angeles River Watershed Monitoring Program workgroup members who participated in the success of this program and report:

Geremew Amenu, Ph.D., *Los Angeles County, Department of Public Works*
Mike Antos, M.S., *Council for Watershed Health*
Stan Asato, M.S., *City of Los Angeles, Environmental Monitoring Division*
Brock Bernstein, Ph.D., *Brock Bernstein Consultants*
Seth Carr, *City of Los Angeles, Regulatory Affairs Division*
Alvin Cruz, *City of Burbank, Public Works Department*
Mas Dojiri, Ph.D., *City of Los Angeles, Environmental Monitoring Division*
Kenneth Franklin, *City of Los Angeles, Environmental Monitoring Division*
Gerry Greene, D.Env., *California Watershed Engineering Corp.*
Emiko Innes, M.S., *Los Angeles County, Department of Public Works*
Chris Lopez, *Los Angeles County, Department of Public Works*
Michael Lyons, M.S., *Los Angeles Regional Water Quality Control Board*
Raphael Mazor, Ph.D., *Southern California Coastal Water Research Project*
Gerry McGowen, Ph.D., *City of Los Angeles, Environmental Monitoring Division*
Karin Patrick, M.S., *Aquatic Bioassay and Consulting Laboratories inc.*
Eric Stein, D.Env., *Southern California Coastal Water Research Project*
Bob Wu, *Caltrans*
Charlie Yu, M.S., *City of Los Angeles, Watershed Protection Division*

Special thanks to Joy Turlo for her early work on this report; Jessie Welcomer, Pomona College student intern; and additional staff from City of Los Angeles Watershed Protection Division and Enviromental Monitoring Division; Aquatic Bioassay and Consulting Laboratories, Inc. field and laboratory staff, and Jason Casanova, Council for Watershed Health.

Cover photos (clockwise from top): 1) Kayaking on the Los Angeles River above Sepulveda Basin, 2) Horseback riding near Hansen Dam, 3) Swimming at Hermit Falls in the Angeles National Forest, 4) Fishing on the Los Angeles River in Long Beach (Photo courtesy of William Preston Bowling).

PURPOSE OF THE REPORT

The Los Angeles Watershed is a dynamic system that is undergoing constant change and a regular, recurring program of monitoring allows us to better understand and respond to changes. The Los Angeles River Watershed Monitoring Program (LARWMP), which is the basis for this report, is a collaborative effort to assess the health of the Los Angeles Watershed from a regional perspective. The motivation for this program came from the Los Angeles Regional Water Quality Control Board (Los Angeles Water Board or LARWQCB) using a unique permit condition for the Cities of Los Angeles and Burbank that initiated a monitoring program designed to increase awareness of issues at the watershed scale through the improved coordination and integration of monitoring efforts.

The Council for Watershed Health[1] manages the LARWMP on behalf of stakeholders, including the major permittees, regulatory and management agencies, and community and conservation groups in the region. The intent of this and subsequent reports is to describe current conditions and trends of the Los Angeles River Watershed through addressing the following five questions:

1. What is the condition of streams in the watershed?

2. Are conditions at areas of unique interest getting better or worse?

3. Are receiving waters near discharges meeting water quality objectives?

4. Is it safe to swim?

5. Are locally caught fish safe to eat?

This is the first time the watershed has been comprehensively assessed using multiple indicators. The results presented in this report will assist watershed managers and other interested persons to identify areas of concern and to prioritize management actions. The detailed assessments, methods, and quality assurance for this program can be found in the individual Los Angeles River Watershed Monitoring Program annual reports from 2008 through 2012.[2] The LARWMP monitoring efforts will continue in future years and the State of the Watershed Report will be issued every five years to reflect new data and findings.

---

1   Originally formed as The Los Angeles & San Gabriel Rivers Watershed Council; the name changed in July 2011.

2   http://watershedhealth.org/programsandprojects/watershedmonitoring.aspx

EXECUTIVE SUMMARY

The Cities of Los Angeles and Burbank and their partners envision a healthy, sustainable Los Angeles River Watershed that meets the water quality, water supply, flood management, recreational and habitat needs of its human and biological communities. With 1,400 miles of streams from the San Gabriel Mountains to the Pacific Ocean, the Los Angeles River Watershed supports a population of more than 4.5 million people and countless plants and animals. We need to understand the watershed's overall health and the major stressors in order to ensure sustainability and resilience of this region.

THE LOS ANGELES RIVER WATERSHED MONITORING PROGRAM

In 2007, local, state, and federal stakeholders formed the Los Angeles River Watershed Monitoring Program (LARWMP) to provide managers and the public with a more complete picture of conditions and trends in the Los Angeles River watershed. The objectives are to develop a watershed-scale understanding of the status of surface waters and improve the coordination and integration of monitoring efforts for both regulatory compliance and ambient watershed condition. The sampling design integrates both random and fixed sites to provide a more complete picture of the watershed as a whole. To determine the overall health of the watershed, we designed a monitoring framework to address five key questions. Prior to this program, little was known about the answers to these questions:

1. What is the condition of streams in the watershed?

2. Are conditions at areas of unique interest getting better or worse?

3. Are receiving waters near discharges meeting water quality objectives?

4. Is it safe to swim?

5. Are locally caught fish safe to eat?

The "State of the Los Angeles River Watershed" report summarizes the results from the first five years of monitoring.

CLIMATE CHANGE AND FIRE

Three previously sampled random sites were burned in the **2009 Station Fire** providing a unique opportunity to monitor the post-fire decline and recovery of benthic macroinvertebrate (BMI) communities and riparian habitat. LARWMP results show the decline in BMI communities , as measured by the IBI, after the fire which have not recovered to pre-fire condition at the time of this report. Some of this decline can be attributed to the destruction of riparian habitat. This habitat is currently recovering at burned sites and continued monitoring will follow the rate and extent of post-fire recovery of plant and animal communities.

QUESTION 1: WHAT IS THE CONDITION OF STREAMS IN THE WATERSHED?

Over the first five years of the program, we sampled fifty randomly placed sites in three watershed subregions of the Los Angeles River Watershed. These sites represented the natural portions of the upper watershed, the effluent-dominated reaches of the mainstem, and the urban tributaries of the lower watershed. Multiple indicators were used to evaluate the condition of streams, including the biological community condition as measured by the Southern California Index of Biological Integrity (IBI), the presence of chemicals of concern, toxicity, and conditions of the physical habitat. **As expected, there was a strong negative relationship between the biological condition of the stream and alteration of the physical habitat.**

During the 5-year monitoring period, benthic macroinvertebrate communities in the more natural streams of the upper watershed had the highest IBI scores. In contrast, we found more degradation in the biological communities in the lower urban and effluent-dominated reaches, as evidenced by lower IBI scores, less diversity in feeding strategies, and the dominance of organisms that were more tolerant of pollution. Future monitoring will focus on improving our understanding of linkages between the condition of the biological communities to the physical, chemical, and toxicological stressors.

QUESTION 2: ARE CONDITIONS AT AREAS OF UNIQUE INTEREST GETTING BETTER OR WORSE?

Assessing how habitat conditions might be changing over time can tell us when restorative or protective measures are needed. We annually monitor the same sites ("fixed sites"), such as river confluences and high value habitats, for changes in conditions, trends, trends due to climate change, and to support other monitoring programs. We collected chemistry, toxicity, biological, and physical habitat data from 16 unique areas of special concern. Among these unique sites were four confluences, the Los Angeles River estuary and nine wetland areas. Since monitoring commenced in the second program year at these locations, four years of data is presented in this report.

At the concrete-lined channels of the four confluences, there were no discernible trends in habitat and biological communities at this time; however, they are degraded relative to the more natural sites in the upper watershed and remained so throughout the survey. Survival (acute) toxicity was not detected at the four confluences during the four-year period. However, reproductive (chronic) toxicity was found at the confluence of the Arroyo Seco with the Los Angeles River.

We assessed the conditions in the Los Angeles River Estuary using sediment chemistry, toxicity and benthic disturbance indicators outlined in the California Sediment Quality Objectives policy. Sediment quality in the estuary was highly variable across the four-year period ranging from 'clearly impacted' to 'unimpacted.'

At the nine wetland sites, we found little change in habitat conditions with better habitat quality overall at upper watershed sites compared to lower watershed sites. We will conduct monitoring of these sites every three to four years going forward, acknowledging the relative stability of conditions.

QUESTION 3: ARE RECEIVING WATERS NEAR DISCHARGES MEETING WATER QUALITY OBJECTIVES?

Stewards of receiving waters need to know the potential impact from known point source discharges in the watershed. We focused on the impact of effluents from three publicly-owned treatment works (POTWs): City of Los Angeles Donald C. Tillman Water Reclamation Plant, City of Burbank Water Reclamation Plant, Cities of Los Angeles and Glendale Water Reclamation Plant.

The POTWs have reduced the levels of bacteria and other pollutants in the River as they are diluted by tertiary-treated effluent. Concentrations of dissolved metals, fecal indicator bacteria, and suspended solids in the effluents are often lower than concentrations in Los Angeles River. The concentration of nitrogenous compounds such as nitrate and ammonia were higher downstream of the discharges; however, they did not exceed water quality objectives. Going forward, we will continue to monitor and evaluate the impact of industrial discharges on water quality in the Los Angeles River.

QUESTION 4: Is it safe to swim?

Swim sites in the Angeles National Forest that are heavily used by the public include Hermit Falls, Eaton Canyon and Sturtevant Falls. During the summer months of 2009 through 2012, a total of 535 samples were collected and analyzed for *E. coli*. **Bacterial levels were elevated at these sites on weekends and holidays when human visitation was the highest.** A study to determine the sources of the bacteria and provide guidance on health impacts, including recommendations for best management practices to reduce bacterial contamination at these swim sites, is currently being conducted.

QUESTION 5: Are locally caught fish safe to eat?

In order to protect the public from the potential risk of eating contaminated fish, managers need data on the levels of unsafe contaminants in fish tissues. We collected and analyzed 96 fish between 2009 and 2012 for four contaminants: mercury, selenium, total DDTs, and total PCBs.

Fish in the Los Angeles River Watershed had lower concentrations of mercury and comparable concentrations of selenium, DDTs, and PCBs compared to fish from other parts of California. Mercury concentrations in largemouth bass measured in the Los Angeles River Watershed were significantly lower than that of those measured nationwide. However, based on these results and recommendations from the California Office of Health and Hazard Assessment, **people are cautioned to not consume largemouth bass caught at Legg Lakes, and limit consumption from Peck Road Park Lake to one meal per week.**

FISH CONSUMPTION FINDINGS

- Three fish species (tilapia, redear sunfish, and bluegill) did not exceed consumption thresholds during the four-year period.

- Largemouth bass and common carp caught in Hansen Lake, Legg Lakes, and Peck Road Water Conservation Park had elevated mercury concentrations to levels that should not be consumed by humans.

- Largemouth bass and common carp in Echo Park Lake, and John Ford Lake contained PCBs at concentrations suggesting that their consumption be limited to one or two meals per week.

Fly fishing on the Los Angeles River near Elysian Valley
(Photo courtesy of William Preston Bowling)

TABLE OF CONTENTS

List of Figures

LIST OF TABLES

LIST OF ACRONYMS

ATL	Advisory Tissue Levels
BMI	Benthic Macroinvertebrates
BWRP	City of Burbank Water Reclamation Plant
CFP	California Floristic Province
CFS	Cubic Feet per Second
CMAP	California's Monitoring and Assessment Program
CRAM	California Rapid Assessment Method
CREST	Cleaner Rivers through Effective Stakeholder-led TMDLs
CSCI	California Stream Condition Index
CTR	California Toxics Rule
CWA	Clean Water Act
DCTWRP	Donald C. Tillman Water Reclamation Plant (City of Los Angeles)
DDT	Dichlorodiphenyltrichloroethane
EMAP	Environmental Monitoring and Assessment Program
EPA	Environmental Protection Agency
FIB	Fecal Indicator Bacteria
FoLAR	Friends of the Los Angeles River
IBI	Index of Biological Integrity (Southern California)
LACC	Los Angeles Conservation Corps
LACDPW	Los Angeles County Department of Public Works
LACFCD	Los Angeles County Flood Control District
LADWP	Los Angeles Department of Water and Power
LAGWRP	City of Los Angeles Glendale Water Reclamation Plant
LARC	Los Angeles Regional Collaborative for Climate Action
LARWMP	Los Angeles River Watershed Monitoring Program
LARWQCB	Los Angeles Regional Water Quality Control Board

LASGRWC	Los Angeles and San Gabriel Rivers Watershed Council
MGD	Million Gallons per Day
MLOE	Multiple Lines of Evidence
MPN	Most Probable Number
MRCA	Mountains Recreation and Conservation Authority
MS4	Municipal Separate Storm Sewer Systems
NDN	Nitrification/Dentrification
NPDES	National Pollutant Discharge Elimination System
OEHHA	Office of Environmental Health Hazard Assessment (State of California)
PBDE	Polybrominated Diphenyl Ethers
PCB	Polychlorinated biphenyls
POTF	Publicly Owned Treatment Facilities
POTW	Publicly Owned Treatment Works
PPT	Parts Per Trillion
PSA	Perennial Streams Assessment
QAPP	Quality Assurance Project Plan
SCAG	Southern California Association of Governments
SCCWRP	Southern California Coastal Water Research Project
SMC	Stormwater Monitoring Coalition
SQO	Sediment Quality Objectives
SWAMP	Surface Water Ambient Monitoring Program
TMDL	Total Maximum Daily Load
TSS	Total Suspended Solids
USACE	United States Army Corps of Engineers
USFS	United States Forest Service
USGS	United States Geological Survey
VOC	Volatile Organic Compounds
WQO	Water Quality Objectives
WRP	Water Reclamation Plants

FIGURE 1. The Los Angeles River Watershed

CHAPTER 1

INTRODUCTION TO THE LOS ANGELES RIVER WATERSHED

A reinvigorated watershed, one that offers clean water, reliable local water supplies, restored native habitats, ample parks, open spaces, recreation opportunities, and integrated flood management, is a watershed that contributes to environmental health, social equity, and economic vitality.

When we try to pick out anything by itself, we find that it is bound fast by a thousand invisible cords - that cannot be broken - to everything in the universe.

- John Muir, 1869

Water quality concerns and the need for a watershed-wide monitoring program are better understood when the watershed is viewed not only as a hydrologic system whose surface waters drain to a single water body, but also as the product of its geophysical processes, biological conditions, cultural and historical context, socioeconomic patterns, and regulatory framework. Current and historical anthropogenic activities including land use practices, industrial development, and modifications to the natural hydrology, have dramatically impacted the nation's watersheds, with consequences for both human and ecological communities. This is particularly true in the Los Angeles River Watershed, home to approximately 4.5 million people[1], and providing habitat for countless native flora and fauna, including endangered species found nowhere else on earth.

Watershed- wide monitoring provides an appropriate scale to accurately assess the impacts of the aforementioned anthropogenic activities on the condition of surface waters in the Los Angeles River Watershed. Topography, geology, soil types, climate, and hydrology, including hydrologic modifications, are factors influencing runoff behavior, and ultimately the introduction of nonpoint source pollutants to the receiving waters.

1.1   ENVIRONMENTAL SETTING

The boundaries of the Los Angeles River Watershed encompass 834 square miles stretching from the San Gabriel Mountains on the northern end of the Los Angeles Basin to the Pacific Ocean. With straightening through channelization, the mainstem of the river measures 51 miles, with the first 32 miles within the City of Los Angeles. The watershed is shaped roughly like a large comma, stretching from the western edge in the Santa Susana and Simi Hills and curving southward around the intrusion of the Santa Monica Mountains to discharge into the Pacific Ocean at Long Beach Harbor in San Pedro Bay *(Figure 1)*.

The topography of the Los Angeles River Watershed is dramatic, dropping from 7,103 feet in the northwestern San Gabriel Mountains to sea level over 51 miles. This corresponds to a drop of 31 feet per mile. For comparison, the Mississippi River is 2,348 miles long and drops approximately 1 foot per mile. The deeply incised, mountain slopes are as steep as 65-70% and are some of the steepest in the world. The most recognizable peak in the watershed, although not the highest, is Mount Wilson, home to the Mount Wilson Observatory and the 100-inch Hooker telescope, dating from 1917 *(Figure 2)*.

The Verdugo Mountains and the San Rafael Hills lie between the eastern edge of the San Fernando Valley and the San Gabriel Mountains. Verdugo Peak, at 3,126 feet, is the highest point in these small ranges and lies entirely within the watershed. To the southeast lies the San Gabriel Valley, the western portion of which is within the Los Angeles River Watershed. Elevations in the mountain-rimmed San Fernando Valley range from 3,747 feet in the north against the Santa Susana Mountains to 1,965 feet in the Santa Monica Mountains. South of the Elysian Hills the coastal plain slopes gently southward with elevations dropping from about 300 feet to sea level over a distance of 20 miles.

To understand the substantial differences in hydrology, as well as habitat and water quality, the LARWMP identifies three hydrologically distinct watershed subregions, as well as the estuary.

---

1   2010 Census Data

FIGURE 2. Topography of the Los Angeles River Watershed *(watershed boundary outlined in blue)*

These subregions are defined:

- Upper Watershed streams dominated by **natural** flows

- The Los Angeles River mainstem (including the Burbank Western Channel) dominated by treatment plant **effluent** flows

- Tributaries in the middle and lower watershed dominated by **urban** runoff

- The intertidal Los Angeles River **estuary**

NATURAL

Approximately one-third (272 square miles) of the Los Angeles River Watershed is in the boundary of the Angeles National Forest and largely managed by the United States Forest Service (USFS). This includes the western portion of the San Gabriel Mountains, the Santa Susana Mountains, the Verdugo Hills, and the northern slope of the Santa Monica Mountains. Big Tujunga Creek is the largest natural perennial stream in the upper watershed.

EFFLUENT

Tertiary-treated effluents from three publicly-owned treatment works (POTWs) dominate dry-weather flows in the river on the coastal plain. Their treatment capacities range from 9 million gallons per day (MGD) for the Burbank Water Reclamation Plant (WRP) to 20

MGD and 80 MGD for the Los Angeles Glendale and Los Angeles D.C. Tillman Water Reclamation Plants, respectively *(Table 1)*. Las Virgenes Municipal Water District's Tapia Plant is permitted to discharge 2 MGD to the Los Angeles River at certain times of the year, but generally discharges much less. These facilities produce recycled water for landscape irrigation and industrial processes. The rest is discharged to surface waters, with the Tillman Plant's discharge first being used for recreation enhancement in Lake Balboa, Wildlife Lake, and Japanese Garden Lake, before flowing into the river.

TABLE 1. POTW Discharges to the Los Angeles River, their design capacities and recycled water production (million gallons per day)

POTW DISCHARGER	DATE BUILT	DESIGN DISCHARGE CAPACITY	RECYCLED WATER PRODUCTION
Las Virgenes Municipal Water District Tapia Plant Waste Water Reclamation Plant (WWRP)	1999-outfall to LA River	2	0
City of Burbank WRP	1966	9	1
City of Los Angeles-Glendale WRP	1976	20	4.5
City of Los Angeles- Tillman WRP	1984	80	26

HISTORICAL HYDROLOGY

Prior to development, the Los Angeles River system was typical of other streams in the southwest. Water and debris from the mountains spread freely across the expansive alluvial plain of the lower watershed. The perennial flow in the Los Angeles River historically originated in large part as rising groundwater from the San Fernando Valley ground water basin. The basin is tilted southwards and overflow waters feed the Los Angeles River as it runs through the southern side of the valley. The river's channel was broad and often shifted location within the flood plain with the high sediment loads; the mouth of the river moved frequently between Long Beach and Ballona Creek. Between 1815 and 1825, the river turned southwest after leaving the Glendale Narrows, where it joined Ballona Creek and discharged into Santa Monica Bay in present-day Marina del Rey (LACDPW, 2006). During a catastrophic flash flood in 1825, its course was diverted again close to its present one, flowing due south just east of present-day downtown Los Angeles and discharging into San Pedro Bay. At this time, the coastal plain was a network of creeks, springs, lakes, and wetlands, only remnants of which still remain today.

FLOOD CONTROL

Following the damaging flood of 1914 and the creation of the Los Angeles County Flood Control District in 1915, a program of flood control and water conservation was initiated in the County. Local residents supported this effort through voter-approved storm drain bond issues in 1952, 1958, 1964, and 1970 for a total of over $900 million (LACDPW, 1996). The County Board of Supervisors approved an additional $200 million bond issue in 1993. In the Los Angeles River Watershed this funded the construction of several dams. The Los Angeles County Flood Control District constructed three major dams, which were completed between 1920 and 1931: Pacoima, Big Tujunga, and Devil's Gate. In the Rio Hondo drainage area, several dams were constructed including Eaton Wash, Sierra Madre, Santa Anita, and Sawpit. As the need outstripped the ability to fund further flood control efforts, the federal government stepped in during the Great Depression. The US Army Corps of Engineers constructed three major dams between 1940 and 1954: Hansen Dam, Sepulveda Dam, and Lopez Dam.

The concrete sections of the Los Angeles River were constructed between the late 1930's and the 1950's in a trapezoidal or rectangular configuration to minimize the costly acquisition of rights-of-way. Most of the channel was lined in concrete to prevent erosion of the native soils and the system was designed to be a low maintenance, durable way to move flood waters through the coastal plain. Three significant portions of the river, however, exist in a seminatural or soft-bottom state: Sepulveda Basin, Glendale Narrows, and the intertidal estuary below Willow Street. In contrast to the concrete-lined portions, today these areas support a wide variety of habitat and wildlife.

Photos: (above) 1850 Los Angeles River through downtown - photo courtesy of Historical Photo Collection of the Department of Water and Power, City of Los Angeles; (above right) present day Glendale Narrows

URBAN

Approximately 4.5 million people live in the highly urbanized middle and lower watershed. The lower watershed is defined as the major tributaries to the Los Angeles River downstream of Sepulveda dam and includes the lower Tujunga Wash, Burbank Western Channel, Verdugo Wash, Arroyo Seco, Rio Hondo, and Compton Creek *(Figure 3)*. During the dry times of the year, flows in these tributaries are dominated by urban runoff that can contain pollutants such as trash, human and animal waste, automobile fluids, industrial pollutants, fertilizers, and pesticides.

LOS ANGELES RIVER ESTUARY

The intertidal Los Angeles River estuary connects the Los Angeles River to San Pedro Bay. The estuary begins where the concrete-lined river ends near Willow Street and flows to Queensway Bay before entering San Pedro Bay. The banks of the soft-bottom estuary are stabilized with rock rip rap. During high tide, the estuary receives most of its flow from either the Los Angeles River or San Pedro Bay. A relatively small area along either bank drains directly to the estuary (approximately 6,000 acres in total land area). Land use in the area is largely residential and commercial (USEPA, 2012).

1.1.1   GEOLOGY AND SOILS

Mountain ranges within or partially within the watershed are part of the Transverse Ranges, so named because they are east-west trending, running counter to the north-south orientation of most ranges in California. These mountain ranges are a work in progress and a product of tectonic activity.

The ranges are among the youngest and fastest rising mountains on earth with some portions currently ascending 5 to 10 millimeters per year (DeCourten, 2010). Uplift is partially counteracted by debris flows and rock falls, expedited by the steepness of the mountain slopes and aided and abetted by fire, which denudes the landscape of stabilizing brush, and intense rain storms, which lubricate the slide. Uplift rates in the Transverse Ranges are estimated to be roughly 7.6 meters per 1,000 years, while the rate of erosion is closer to 2.3 meters per 1,000 years (Wohlgemuth, 2006).

The San Gabriel Mountains are generally composed of Mesozoic and older igneous and metamorphic rock. The Santa Susana Mountains are composed mainly of Miocene to Pleistocene marine and non-marine

FIGURE 3. Drainage area of subwatersheds (Total: 834 mi²)

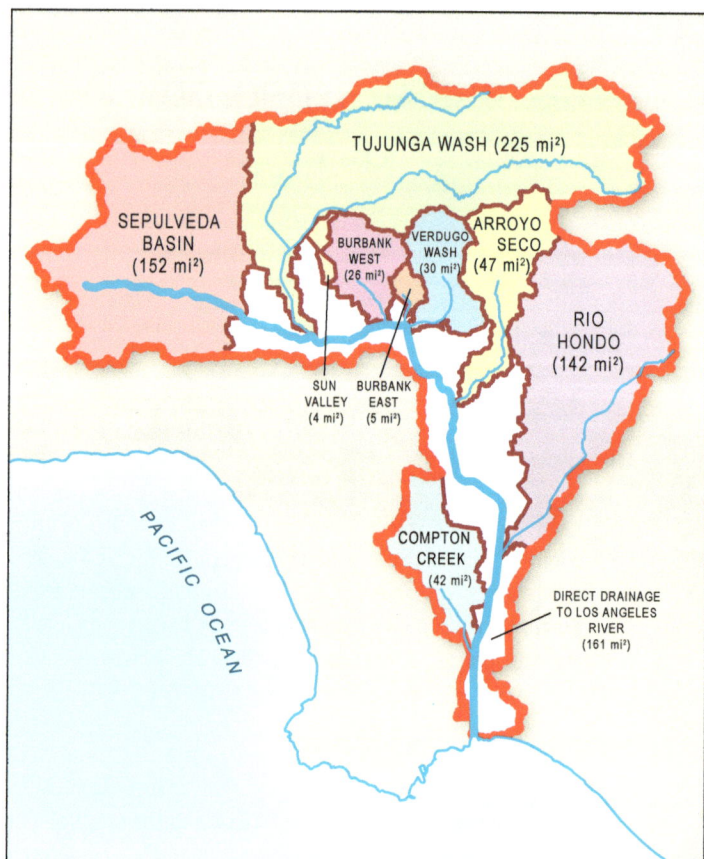

sedimentary rock. The Santa Monica Mountains are composed mainly of Cretaceous to Miocene sedimentary and volcanic rock.

It is this topology and geology that created the rich alluvial deposits that originally attracted farmers to the San Gabriel Valley, the eastern portion of the San Fernando Valley, and a large part of the coastal plain. Closest to the mountains, coarse gravel predominates while the granularity of the deposits diminishes in size with distance from the San Gabriel Mountains, graduating down to sand, silt, and clay. In the central and western portions of the San Fernando Valley, the deposits are fine-grained materials resulting from the erosion of shale, sandstone, and clay, with most of the material having been deposited by streams entering the valley from the southern slopes of the Santa Susana Mountains *(Figure 4)*.

1.1.2   CLIMATE

The watershed is situated in a Mediterranean climate zone, characterized by warm, dry summers and cool, wet winters. It is this climate that is largely responsible for

FIGURE 4. Geologic map of the Los Angeles River Watershed

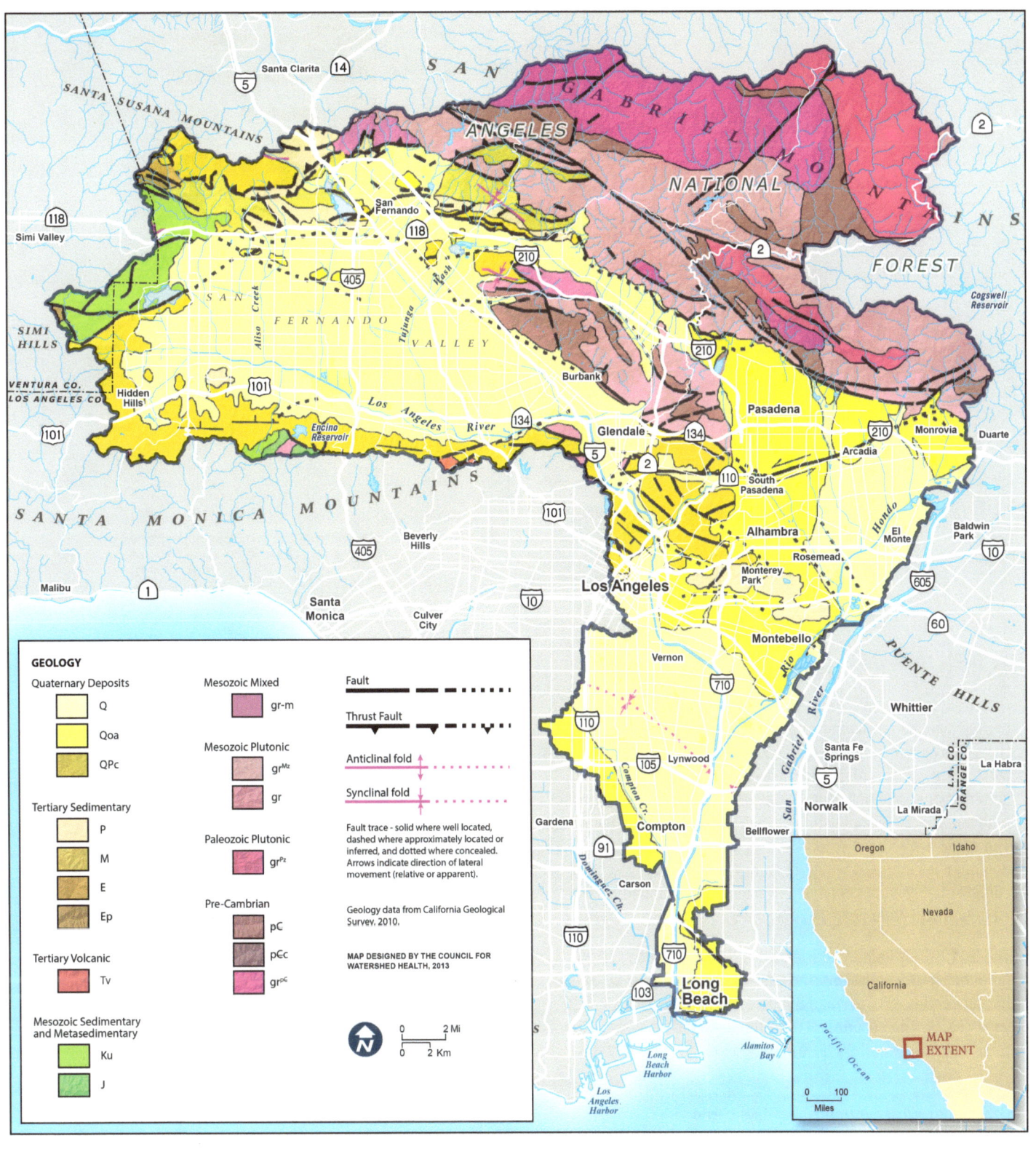

GEOLOGY

Quaternary Deposits
Q
Qoa
QPc

Tertiary Sedimentary
P
M
E
Ep

Tertiary Volcanic
Tv

Mesozoic Sedimentary
and Metasedimentary
Ku
J

Mesozoic Mixed
gr-m

Mesozoic Plutonic
grMz
gr

Paleozoic Plutonic
grPz

Pre-Cambrian
pC
pCc
grpC

Fault
Thrust Fault
Anticlinal fold
Synclinal fold

Fault trace - solid where well located,
dashed where approximately located or
inferred, and dotted where concealed.
Arrows indicate direction of lateral
movement (relative or apparent).

Geology data from California Geological
Survey, 2010.

MAP DESIGNED BY THE COUNCIL FOR
WATERSHED HEALTH, 2013

0   2 Mi
0   2 Km

MAP EXTENT

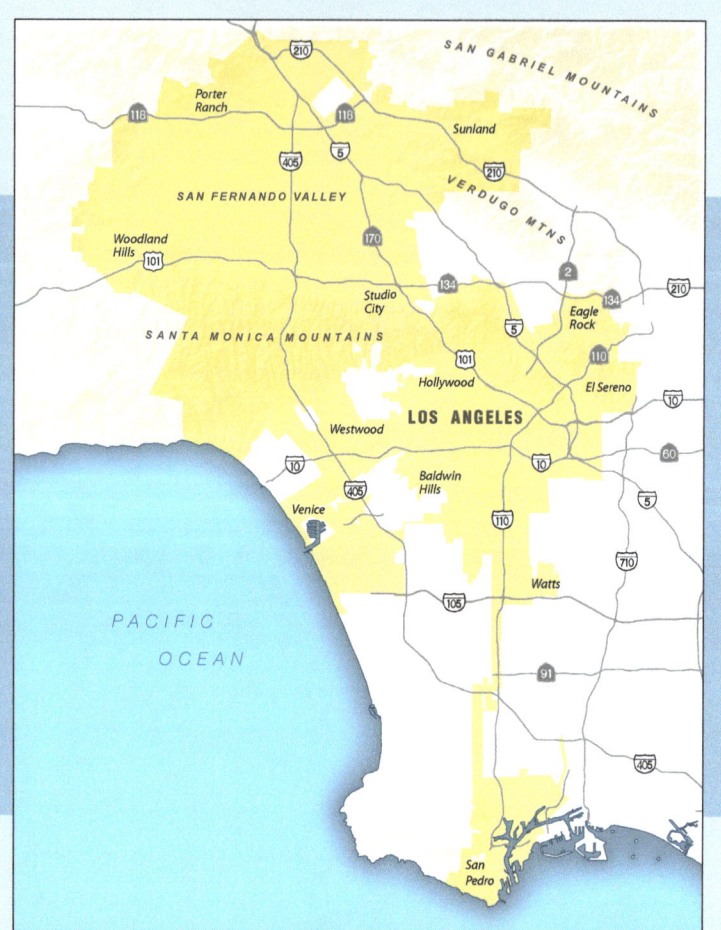

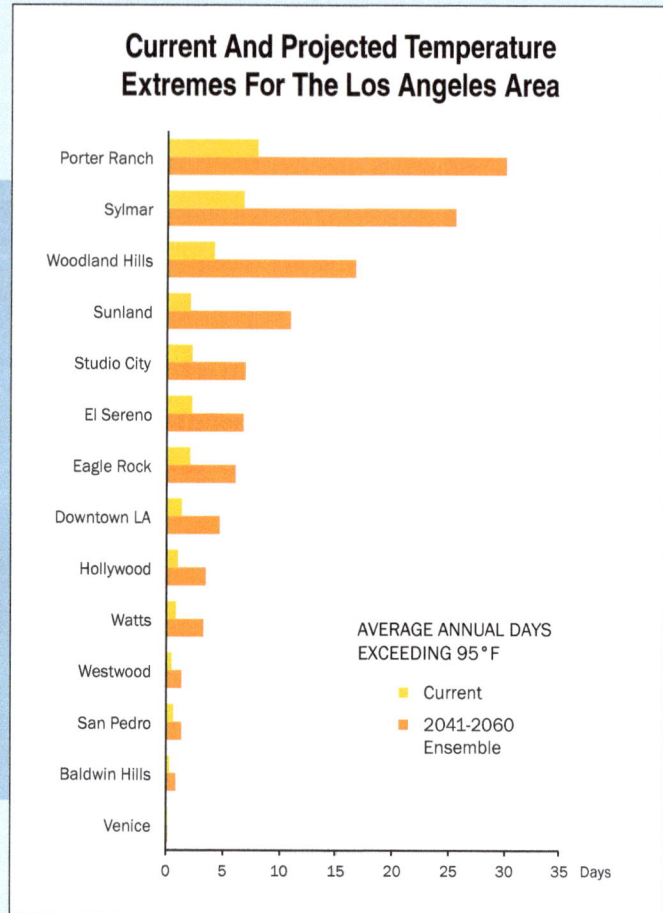

Current And Projected Temperature Extremes For The Los Angeles Area

AVERAGE ANNUAL DAYS EXCEEDING 95°F
- Current
- 2041-2060 Ensemble

(Source: UCLA LARC study, 2012; chart based on the mean/average projected by the 19 climate models.

REGIONAL CLIMATE CHANGE

Climate change impacts over the next century are expected to amplify regional climate variability. The City of Los Angeles, through a partnership with the Los Angeles Regional Collaborative for Climate Action (LARC) and the Department of Energy, commissioned the University of California Los Angeles to conduct a high-resolution climate study to allow for better urban planning. The study benefits the other 43 cities and unincorporated communities in the Los Angeles River Watershed by predicting climate change impacts on a finer scale, neighborhood by neighborhood, to improve local planning.

The 2012 study found that even with efforts to reduce greenhouse gas emissions, the Los Angeles region will be warmer by mid-century, with average annual temperatures rising **4-5 °F** (Hall, 2012). The occurrence of "extreme heat days," days when temperatures exceed 95°F, is expected to increase substantially.

Coastal areas and central L.A. (the areas with the highest population) will see extreme heat days **triple**.

The San Fernando Valley and San Gabriel Valley will see extreme hot days almost **quadruple**.

Desert and mountain areas will see extreme hot days increase by **5 to 6 times**.

Worldwide, the higher temperatures are expected to lead to severe drought, lower soil moisture, and frequent fires, factors which will result in changes in the geographic ranges of plant and animal species as they migrate to higher latitudes or higher elevations. Invasive species, which are typically generalists with a higher tolerance range, will multiply.

Higher water temperatures in lakes and streams are expected to result in a reduction in dissolved oxygen, a change that will stress freshwater fish and invertebrates. Impacts to the Sacramento-San Joaquin Bay Delta and the Colorado River from earlier snowmelt, drought, and other climate related phenomena could cause major water shortages in Southern California. Reduced stream flow in summer may increase the concentration of pollutants through reduced dilution.

Moreover, warming ocean temperatures are expected to result in sea level rise. This will have its greatest impact in coastal areas, causing coastal erosion and seawater intrusion, and possibly putting a strain on water resources in the region.

For more impacts see *http://downloads.globalchange.gov/usimpacts/pdfs/water.pdf*

the settlement of Native Americans and later promoted westward migration and settlement in the Los Angeles region. From medical doctors who sent tuberculosis patients west to recuperate, to the promotion of the region for growing citrus and other crops, the climate enticed people to Los Angeles.

The seasonal variability in precipitation and temperature *(Figure 5)* demonstrate characteristic Mediterranean climate conditions. In contrast, the spatial variation in local climate is largely a result of the topography of the region.

Moisture-laden air from the ocean moves up the slopes of the San Gabriel Mountains, cooling as it rises and creating a barrier that traps moist ocean air against the mountain slopes and partially blocks summer heat from the desert and winter cold from the interior northeast.

Rainfall in the mountains increases with elevation. Altadena, nestled in the San Gabriel Mountain foothills at roughly 1,300 feet, receives an annual average of 22 inches, while Mt. Wilson at 5,712 feet receives 35 inches. Historically the San Gabriel Mountains have experienced high intensity record-breaking storms, during which heavy rainfall occurs over a relatively short period

of time *(Figure 6)*. Average annual rainfall in the valleys and coastal plain diminishes from north to south, with the central San Fernando Valley receiving roughly 17 inches, the Central Basin south of Los Angeles receiving approximately 15 inches, and the City of Long Beach receiving an average of just under 13 inches annually.

Along the same profile from coast to mountain, average annual temperatures vary across the watershed. The greatest seasonal temperature variations were recorded at the Mt. Wilson station, followed by the Van Nuys station in the San Fernando Valley.

1.2 LAND USE

Surface water quality depends greatly on, and reflects differences in, various land uses in the Los Angeles River Watershed. The upper watershed is predominantly forested open space that provides recreational opportunities such as hiking and swimming. In contrast, the highly urbanized middle and lower watershed supports primarily residential and industrial land uses *(Figure 7 and Table 2)*. 2010 census data assisted in delineating the upper and lower watershed. Since the 1990 Southern California Association of Goverments

FIGURE 5. Mean annual rainfall and temperature across the Los Angeles River Watershed (1931-2000).

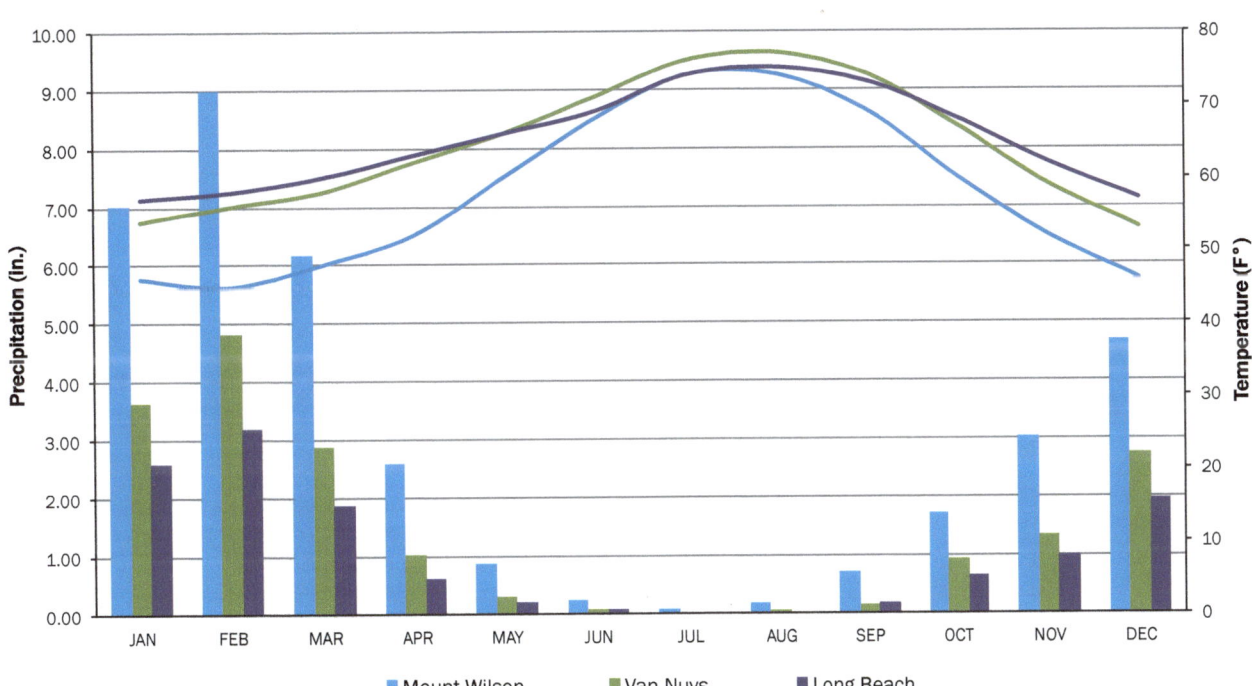

FIGURE 6. Rainfall intensity (24-hr event; 100-yr return period) in the Los Angeles River Watershed.

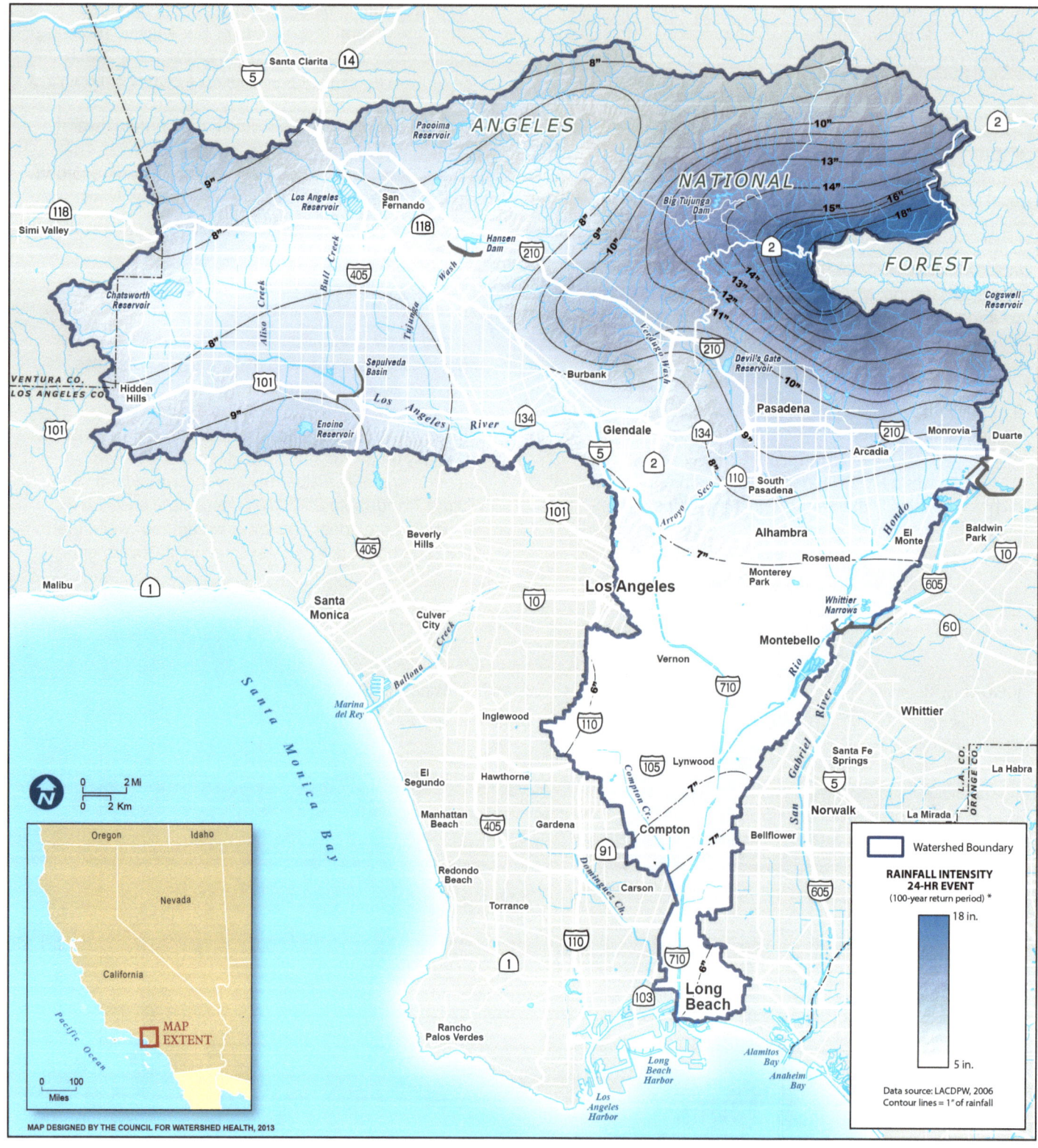

FIGURE 7. Land use in the Los Angeles River Watershed

TABLE 2. Land use in the Los Angeles River Watershed[1]

Location		Residential	Commercial	Industrial	Agriculture	Recreation/Forest	Total
Watershed	Acres	198,518	28,420	74,773	3,153	228,767	533,630
	%	37.20	5.33	14.01	0.59	42.87	100.00
Upper Watershed (Natural)	Acres	671	21	2,242	238	176,142	179,314
	%	0.37	0.01	1.25	0.13	98.23	100.00
Middle and Lower Watershed (Urban and Mainstem)	Acres	197,847	28.399	72,531	2,915	52,625	354,316
	%	55.84	8.02	20.47	0.82	14.85	100.00

1 Land use information was obtained from the 2005 Southern California Association of Governments (SCAG) 2005 Aerial Land Use study. The land use categories shown here were generalized from the categories shown in *Figure 7*. Watershed locations are illustrated in *Figure 1*.

(SCAG) land use study, agricultural and commercial uses have decreased slightly, while industrial uses have increased from 6% to 14%.

1.3 BIODIVERSITY

The Los Angeles River Watershed lies within the California Floristic Province, one of the world's biodiversity hotspots. Biodiversity hotspots are regions that support especially high numbers of endemic species, or species that occur naturally nowhere else on Earth. The concept was defined in 1988 by British ecologist Norman Myers to address the dilemma that conservationists face: what areas are the most immediately important for conserving biodiversity? Destruction of habitat is the leading cause of biodiversity loss, but invasive species, pollution, overexploitation, and climate change pose major threats as well.

FLORA

Naturally occurring plant communities of the Los Angeles River Watershed include coastal sage scrub, alluvial fan sage scrub, chaparral, coast live oak woodlands, California walnut woodlands, riparian woodlands, riparian wetlands, and mixed coniferous forest *(Table 3)*. Many of the plant and animal species associated with these plant communities are listed by state and federal agencies as rare, sensitive, threatened, or endangered. Due to the channelization and loss of protective vegetative cover many species have declined or are extirpated. Urbanization has greatly diminished and fragmented the distribution of all of these plant communities. Coastal sage scrub, for example, is an endangered plant community with only 10-15% of its historic range remaining.

FAUNA

Few comprehensive studies summarize the richness of wildlife within the Los Angeles River system. One study from 1993 concluded that although highly urbanized, the Los Angeles River watershed continues to support a variety of wildlife in the habitats that remain (Garrett, 1993). The mountain and foothill areas of the upper watershed support more native species than the coastal plain and many species have federal or state protected status.

Birds in particular thrive in the soft-bottom habitats of the Sepulveda Basin, a 2–mile stretch of Glendale Narrows (a soft- bottom channel between Atwater Village and Elysian Park), and the highly disturbed reach of the Rio Hondo to the confluence with Compton Creek (Bloom et al., 2002). Over 400 bird species have been recorded throughout the watershed.

In contrast, channelization has dramatically impacted native fish populations and four of the seven native species are extirpated. The three remaining species include Santa Ana sucker, arroyo chub, and Santa Ana speckled dace. Habitat destruction has reduced the historic number of mammal, reptile, and amphibian species. Grizzly bears and the western pond turtle are examples of charismatic species that have been extirpated from the watershed.

Photo: (above) Arboreal Salamander courtesy of Micheal Ready

TABLE 3. Plant communities

PLANT COMMUNITY	CURRENT STATUS	LOCATIONS
Coastal sage scrub	Fragmented throughout its range; considered endangered habitat; 10-15% of its historic range remains.	Santa Monica Mountains, Verdugo Mountains, San Gabriel Mountains, Simi Hills, Santa Susana Mountains, Arroyo Seco
Alluvial fan sage scrub	Eliminated from most of its former range.	Big Tujunga Wash
Chaparral	Still abundant within the mountains of Southern California. Fire is an important factor in ecology of this community.	Santa Monica Mountains, Verdugo Mountains, San Gabriel Mountains, Simi Hills, Santa Susana Mountains, Arroyo Seco
Coast live oak woodlands	Impacted by development, overgrazing and, most recently, now threatened by gold-spotted oak borer.	Santa Monica Mountains, Verdugo Mountains, Simi Hills, Santa Susana Mountains, Arroyo Seco, San Gabriel Mountains, south facing foothills
California walnut woodlands	Designated sensitive habitat type by CA Fish and Wildlife. Only 14,332 acres remain.	Santa Monica Mountains, Simi Hills, Santa Susana Mountains, Arroyo Seco, San Gabriel Mountains, south facing foothills, Repetto Hills
Riparian woodlands	Only 3-5% of its historic range remains.	Santa Monica Mountains, Simi Hills, Santa Susana Mountains, Arroyo Seco, San Gabriel Mountains, Big Tujunga, Los Angeles River segments
Riparian wetlands	Eliminated from most of its former range.	Small under-developed patches in soft-bottomed reaches: Compton Creek, Sepulveda Basin, Hansen Dam, Glendale Narrows, Los Angeles River estuary
Mixed coniferous forest	Heavily impacted by forest management practices and recreational use.	Upper Big Tujunga, Arroyo Seco, San Gabriel Mountains

Adapted from Appendix A of the Los Angeles River Master Plan

Nourishing an Expanding Population
HISTORICAL OVERVIEW OF THE

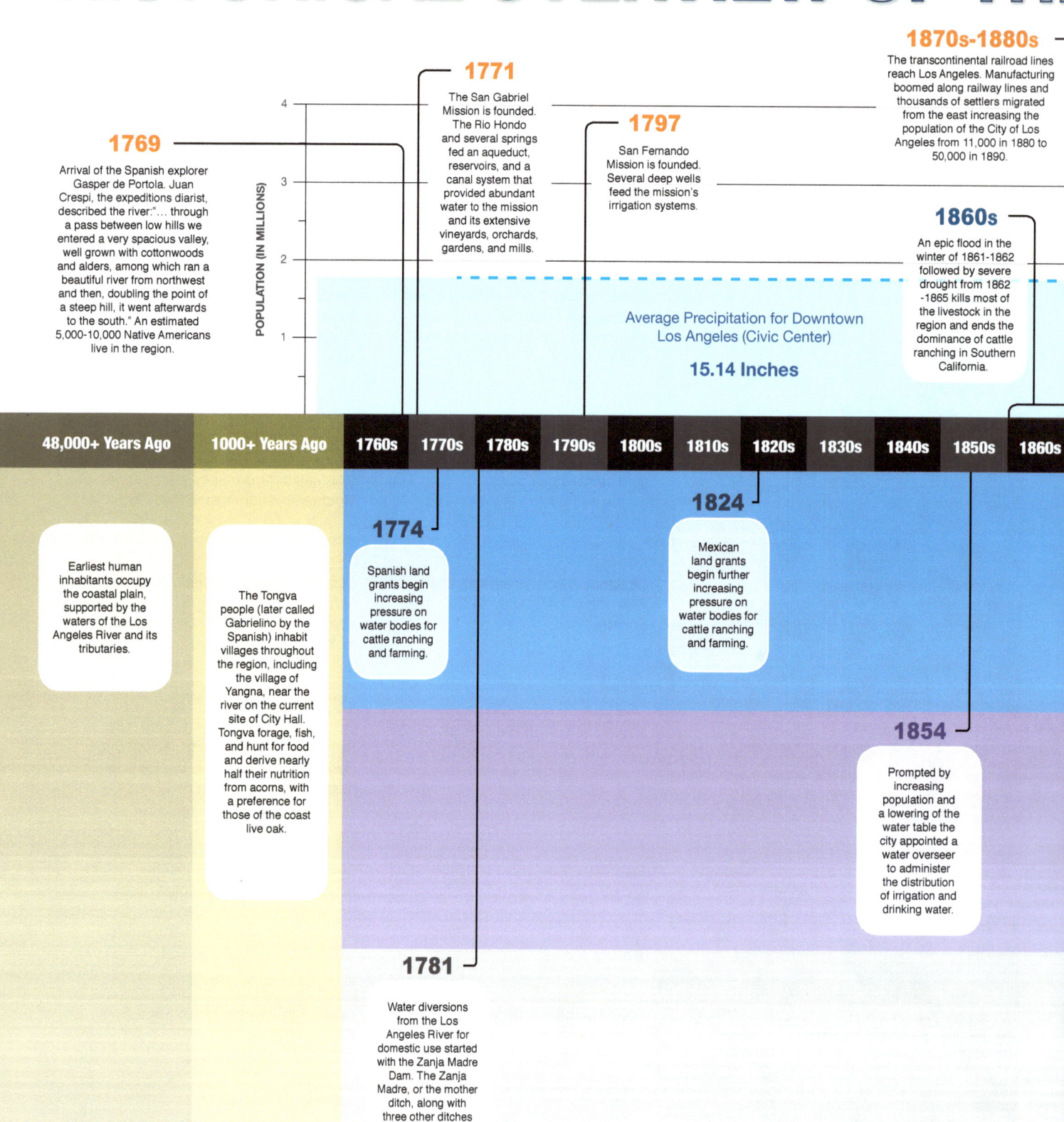

1769

Arrival of the Spanish explorer Gasper de Portola. Juan Crespi, the expeditions diarist, described the river:"… through a pass between low hills we entered a very spacious valley, well grown with cottonwoods and alders, among which ran a beautiful river from northwest and then, doubling the point of a steep hill, it went afterwards to the south." An estimated 5,000-10,000 Native Americans live in the region.

1771

The San Gabriel Mission is founded. The Rio Hondo and several springs fed an aqueduct, reservoirs, and a canal system that provided abundant water to the mission and its extensive vineyards, orchards, gardens, and mills.

1797

San Fernando Mission is founded. Several deep wells feed the mission's irrigation systems.

1870s-1880s

The transcontinental railroad lines reach Los Angeles. Manufacturing boomed along railway lines and thousands of settlers migrated from the east increasing the population of the City of Los Angeles from 11,000 in 1880 to 50,000 in 1890.

1860s

An epic flood in the winter of 1861-1862 followed by severe drought from 1862 -1865 kills most of the livestock in the region and ends the dominance of cattle ranching in Southern California.

POPULATION (IN MILLIONS)

4
3
2
1

Average Precipitation for Downtown
Los Angeles (Civic Center)

15.14 Inches

48,000+ Years Ago	1000+ Years Ago	1760s	1770s	1780s	1790s	1800s	1810s	1820s	1830s	1840s	1850s	1860s

Earliest human inhabitants occupy the coastal plain, supported by the waters of the Los Angeles River and its tributaries.

The Tongva people (later called Gabrielino by the Spanish) inhabit villages throughout the region, including the village of Yangna, near the river on the current site of City Hall. Tongva forage, fish, and hunt for food and derive nearly half their nutrition from acorns, with a preference for those of the coast live oak.

1774

Spanish land grants begin increasing pressure on water bodies for cattle ranching and farming.

1824

Mexican land grants begin further increasing pressure on water bodies for cattle ranching and farming.

1854

Prompted by increasing population and a lowering of the water table the city appointed a water overseer to administer the distribution of irrigation and drinking water.

1781

Water diversions from the Los Angeles River for domestic use started with the Zanja Madre Dam. The Zanja Madre, or the mother ditch, along with three other ditches formed the primary water distribution system in Los Angeles from 1781 to the early 1900s.

LOS ANGELES RIVER

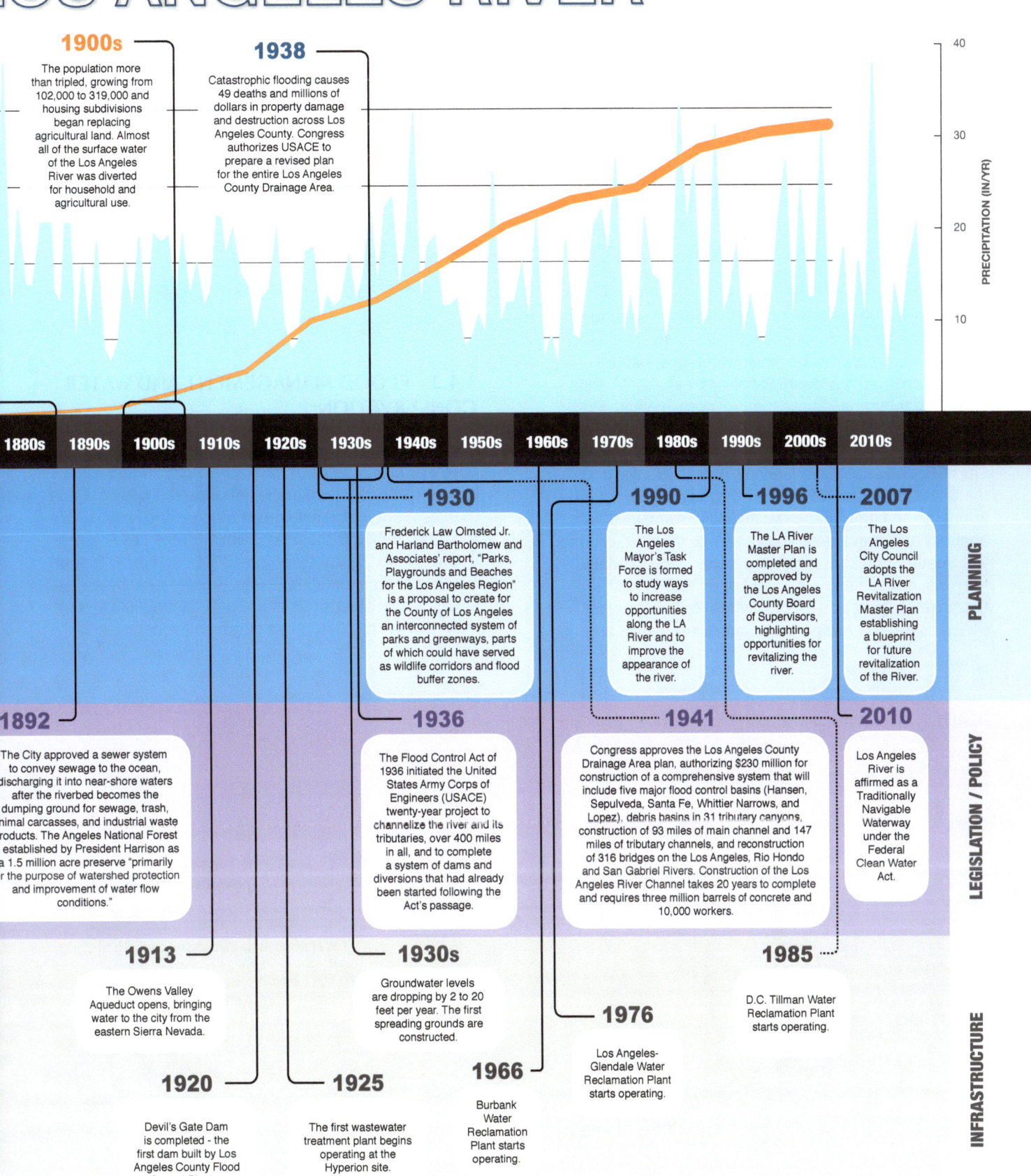

1900s

The population more than tripled, growing from 102,000 to 319,000 and housing subdivisions began replacing agricultural land. Almost all of the surface water of the Los Angeles River was diverted for household and agricultural use.

1938

Catastrophic flooding causes 49 deaths and millions of dollars in property damage and destruction across Los Angeles County. Congress authorizes USACE to prepare a revised plan for the entire Los Angeles County Drainage Area.

PRECIPITATION (IN/YR)

40 — 30 — 20 — 10

| 1880s | 1890s | 1900s | 1910s | 1920s | 1930s | 1940s | 1950s | 1960s | 1970s | 1980s | 1990s | 2000s | 2010s |

PLANNING

1930

Frederick Law Olmsted Jr. and Harland Bartholomew and Associates' report, "Parks, Playgrounds and Beaches for the Los Angeles Region" is a proposal to create for the County of Los Angeles an interconnected system of parks and greenways, parts of which could have served as wildlife corridors and flood buffer zones.

1990

The Los Angeles Mayor's Task Force is formed to study ways to increase opportunities along the LA River and to improve the appearance of the river.

1996

The LA River Master Plan is completed and approved by the Los Angeles County Board of Supervisors, highlighting opportunities for revitalizing the river.

2007

The Los Angeles City Council adopts the LA River Revitalization Master Plan establishing a blueprint for future revitalization of the River.

LEGISLATION / POLICY

1892

The City approved a sewer system to convey sewage to the ocean, discharging it into near-shore waters after the riverbed becomes the dumping ground for sewage, trash, animal carcasses, and industrial waste products. The Angeles National Forest is established by President Harrison as a 1.5 million acre preserve "primarily for the purpose of watershed protection and improvement of water flow conditions."

1936

The Flood Control Act of 1936 initiated the United States Army Corps of Engineers (USACE) twenty-year project to channelize the river and its tributaries, over 400 miles in all, and to complete a system of dams and diversions that had already been started following the Act's passage.

1941

Congress approves the Los Angeles County Drainage Area plan, authorizing $230 million for construction of a comprehensive system that will include five major flood control basins (Hansen, Sepulveda, Santa Fe, Whittier Narrows, and Lopez), debris basins in 31 tributary canyons, construction of 93 miles of main channel and 147 miles of tributary channels, and reconstruction of 316 bridges on the Los Angeles, Rio Hondo and San Gabriel Rivers. Construction of the Los Angeles River Channel takes 20 years to complete and requires three million barrels of concrete and 10,000 workers.

2010

Los Angeles River is affirmed as a Traditionally Navigable Waterway under the Federal Clean Water Act.

INFRASTRUCTURE

1913

The Owens Valley Aqueduct opens, bringing water to the city from the eastern Sierra Nevada.

1930s

Groundwater levels are dropping by 2 to 20 feet per year. The first spreading grounds are constructed.

1985

D.C. Tillman Water Reclamation Plant starts operating.

1976

Los Angeles-Glendale Water Reclamation Plant starts operating.

1920

Devil's Gate Dam is completed - the first dam built by Los Angeles County Flood Control District.

1925

The first wastewater treatment plant begins operating at the Hyperion site.

1966

Burbank Water Reclamation Plant starts operating.

1.4 WATER RESOURCES

1.4.1 SURFACE WATER

Annual stream flows in the Los Angeles River reflect the Mediterranean climate along with flood management and water conservation practices. The average monthly stream flows at five sites located throughout the watershed highlight the seasonal and spatial variability *(Figure 8)*. In general, flows increase downstream in the watershed and flows in the Los Angeles River, Compton Creek, Verdugo Wash, and Burbank Western Channel have steadily increased since the 1940s, as impervious cover has increased.

The typical dry-weather period from May through September is characterized by little or no rainfall and steady flows that range from 1 cubic feet per second (CFS) at the headwaters up to 194 CFS at the confluence with San Pedro Bay. In the upper watershed, natural springs feed Tujunga Wash, Pacoima Wash, Santa Anita and other tributaries above their respective dams. Flow in the lower watershed is sustained by treated effluents from three publicly-owned treatment works (POTWs): the City of Los Angeles' Glendale and Tillman POTWs, and the City of Burbank POTW. The POTWs proportion of total annual stream flow in the Los Angeles River varies both annually and seasonally, and can range from 19% during wet weather to 92% during dry weather. The Glendale Narrows, a seven-mile long soft-bottom section of the Los Angeles River adjacent to Griffith Park, is an area where rising groundwater also contributes significant

dry-weather flows into the river. Historically this rising groundwater ensured that the river had year-round flow (LASGRWC, 2001).

General urban runoff is the source of most of the dry-season flow in many of the tributaries and channels of the lower watershed. Approximately 100 million gallons of runoff from landscape irrigation, car washing, and other inadvertent sources flows through the Los Angeles County storm drain system daily and into the flood control channels, including the Los Angeles River and its tributaries (Sheng, 2009).

The typical wet period spans October through April and flows range from 1.34 CFS at the headwaters up to 1,592 CFS at the estuary. This period is marked by occasional storms and flows during storm events are flashy. Storms can increase runoff volume to 10 billion gallons (Sheng, 2009).

1.4.2 FLOOD MANAGEMENT AND WATER CONSERVATION

Expanding urban development and periodic droughts have increased the need to conserve stormwater runoff behind dams and to recharge groundwater basins. Flood management and water conservation in the Los Angeles River Watershed is the responsibility of the Los Angeles County Flood Control District (LACFCD) and the U.S. Army Corps of Engineers (USACE). The present system is a sophisticated integration of the natural drainage system with heavily engineered hydrologic components.

FIGURE 8. Average monthly stream flow in the Los Angeles River Watershed.

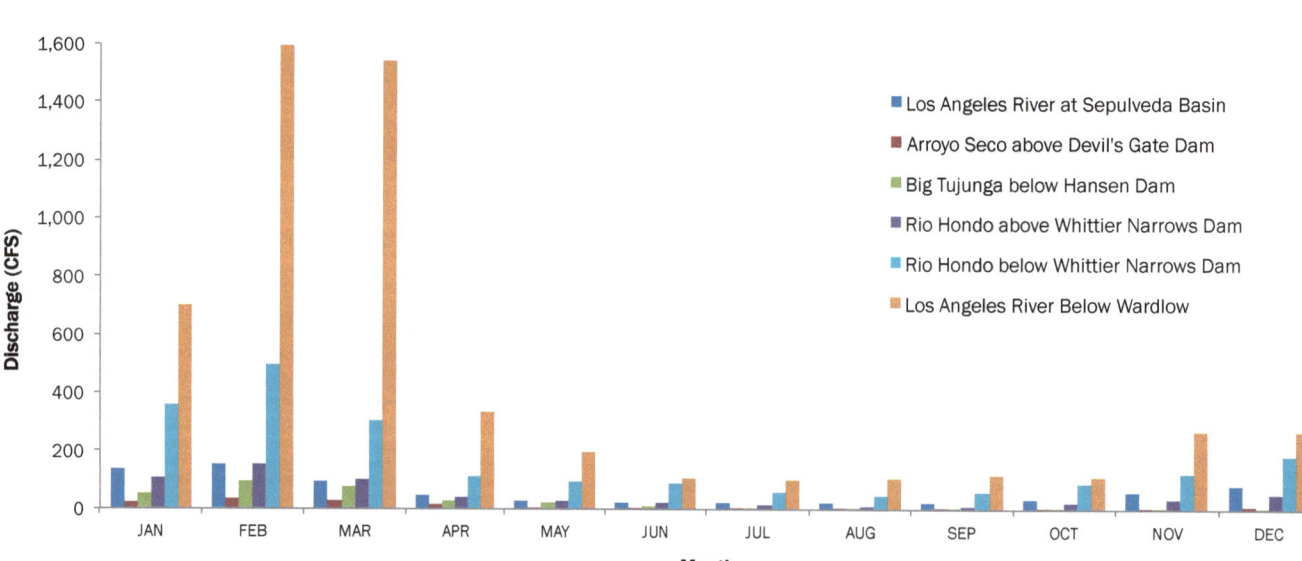

Within the Los Angeles River Watershed, USACE operates and maintains three major flood control reservoirs (Hansen, Sepulveda and Lopez)(*Figure 9 and Table 4*). In addition to providing flood protection, these facilities control upstream debris flows and provide recreational opportunities. The LACFCD operates and maintains three major dams and numerous sediment entrapment basins in the Watershed. Local storm drains and pump stations are maintained by LACFCD, cities, Caltrans and specific homeowner associations (LACDPW, 1996).

FIGURE 9. Flood management in the Los Angeles River Watershed.

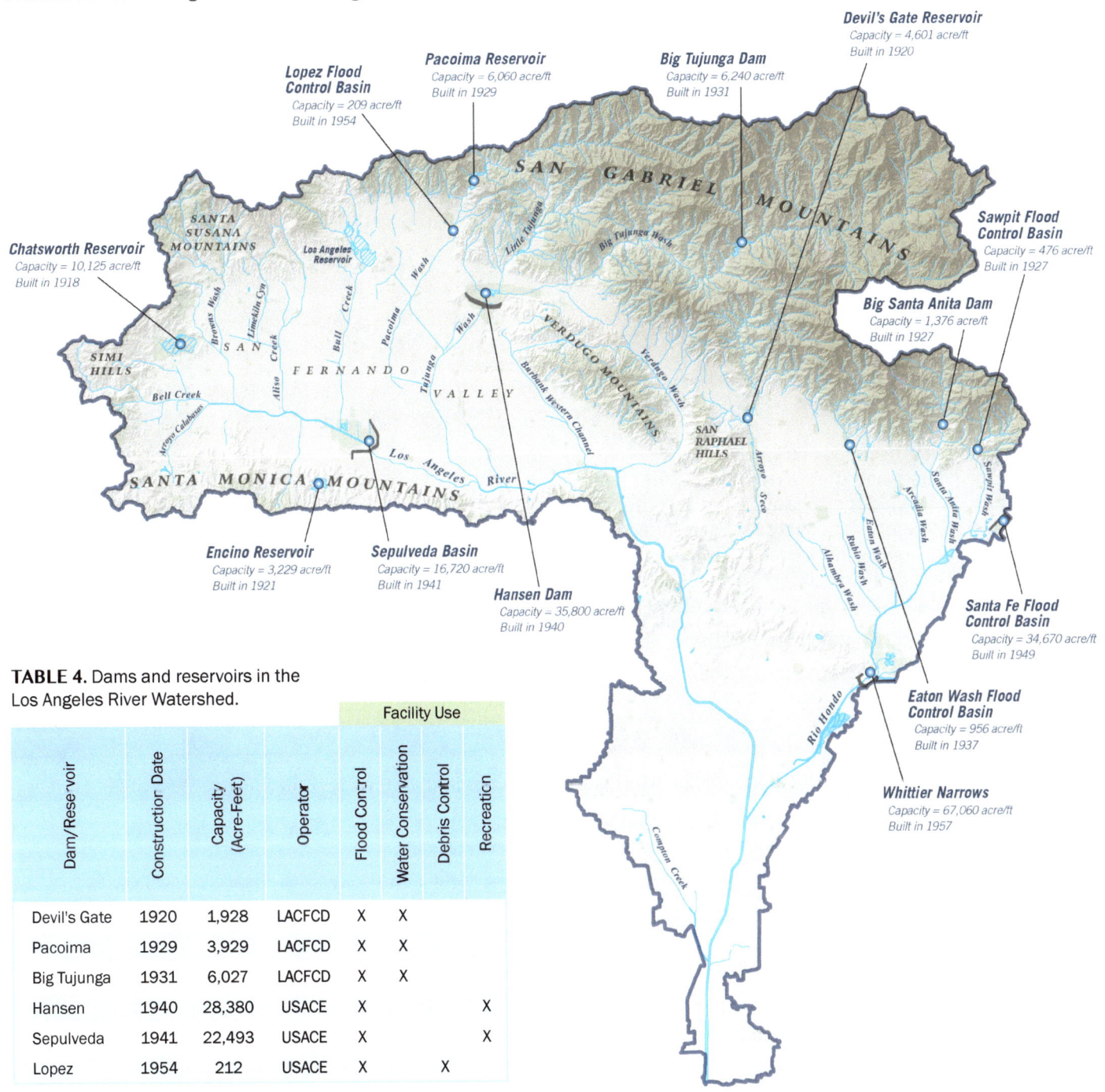

TABLE 4. Dams and reservoirs in the Los Angeles River Watershed.

Dam/Reservoir	Construction Date	Capacity (Acre-Feet)	Operator	Flood Control	Water Conservation	Debris Control	Recreation
				Facility Use			
Devil's Gate	1920	1,928	LACFCD	X	X		
Pacoima	1929	3,929	LACFCD	X	X		
Big Tujunga	1931	6,027	LACFCD	X	X		
Hansen	1940	28,380	USACE	X			X
Sepulveda	1941	22,493	USACE	X			X
Lopez	1954	212	USACE	X	X		

1.4.3 GROUNDWATER

Groundwater accounts for most of the region's local supply of freshwater and a priority of LACFCD is to conserve the maximum amount of stormwater possible to recharge groundwater basins (LACDPW, 1996). The amount of water that is recharged annually is determined by the quality and quantity of stormwater, imported water (from the Colorado River and the Owens Valley) and recycled water available for recharge, capacities of spreading grounds, and the geologic and groundwater conditions. LACFCD, in concert with the Los Angeles Department of Water and Power and the Water Replenishment District of Southern California, operates and maintains 3,361 acres of spreading grounds and soft-bottom channel spreading areas in the County. In contrast to the neighboring San Gabriel River Watershed, much of the Los Angeles River Watershed is underlain with extensive clay layers and the most important spreading basins are in the San Fernando Valley where the underlying soils are permeable (LASGRWC, 2001).

Four major groundwater basins underlie the extent of the Los Angeles River Watershed: San Fernando Basin, Raymond Basin, Main San Gabriel Basin, and the Central Basin. The West Coast Basin underlies a small portion of the watershed to the south *(Table 5 and Figure 10)*.

In general, historical activities and practices have degraded the groundwater quality in the County over the past century. Causes include seepage of fertilizers and pesticides into the subsurface from past agricultural uses, nitrogen and pathogenic bacteria from poorly sited and maintained septic tanks, and various hazardous substances from leaking aboveground and underground storage tanks and industrial operations. Overdraft of groundwater from coastal aquifers in the first half of the 20th Century resulted in not only a decline in groundwater levels, but also the intrusion of seawater into the aquifers.

Tujunga Spreading Grounds (Photo courtesy of Los Angeles Department of Water and Power)

TABLE 5. Groundwater basins[1]

BASIN	AREA (SQ.MI.)	STORAGE CAPACITY (AF)	RECHARGE
San Fernando Basin	226	3.67 million	Spreading grounds downstream of Tujunga, Hansen, and Pacoima dams
Main San Gabriel Basin	167	10.74 million	San Gabriel River and Rio Hondo spreading grounds , Raymond and Chino Basins
Raymond Basin	40	1.45 million	Hahamongna, Eaton, Sierra Madre, and Santa Anita spreading grounds
Central Basin	270	13.8 million	Montebello forebay, Hollywood Basin, Rio Hondo and San Gabriel River spreading grounds, Dominquez Gap Wetlands
West Coast Basin	140	6.5 million	Direct injection and Central Basin lateral flow

1   California. (1968). Planned utilization of ground water basins: Coastal Plain of Los Angeles County. Sacramento.
    California. (2003). California's groundwater. Sacramento, Calif.: Dept. of Water Resources.

FIGURE 10. Groundwater basins of the Los Angeles River Watershed.

TABLE 6. Beneficial uses of water bodies in the Los Angeles River Watershed

USE CATEGORY	ESTUARY	ABOVE ESTUARY
POPULATION USES		Municipal and domestic supply Industrial process supply Groundwater recharge
	Navigation Industrial service supply	
RECREATION AND COMMERCIAL USES	Commercial and sport fishing	
	Non-contact water recreation Water contact recreation	
HABITAT-RELATED USES	Estuarine habitat Marine habitat Migration of aquatic organisms	Warm freshwater habitat Cold freshwater habitat
	Wetland habitat Wildlife habitat Rare, threatened, or endangered species Spawning, reproduction, and/or early development	

1.5 WATER QUALITY

1.5.1 BENEFICIAL USES

The protection of surface water quality in the Los Angeles River Watershed is regulated by the Los Angeles Regional Water Quality Control Board (LARWQCB) through the Basin Plan for the Coastal Watersheds of Los Angeles and Ventura Counties (LARWQCB, 1994). The Basin Plan identifies surface and groundwater bodies, designates applicable beneficial use classifications to each water body (Table 6), establishes general and water body-specific water quality objectives, and suggests an implementation plan for maintaining or restoring the water quality objectives. The Water Boards utilize NPDES permits and Waste Discharge Requirements to limit the discharge of contaminants and protect surface water quality. Each stream segment may have multiple beneficial use designations. Table 6 lists all the beneficial use classifications in the watershed.

1.5.2 PERMITTED DISCHARGES

The LARWQCB controls pollution in the Los Angeles River and some of its tributaries by issuing permits to point source dischargers. NPDES general permits are issued to multiple point source dischargers within specific categories, based on similarity of operations, discharges, required effluent limitations, monitoring requirements, and other factors. This allows a large number of facilities to be covered under a single permit. Of the 117 entities that hold NPDES discharge permits, 101 are covered under general permits and 16 hold individual permits, including Boeing Company Santa Susana Field Laboratory and the Los Angeles Turf Club.

Minor permits cover miscellaneous wastes such as ground water dewatering, recreational lake overflow, swimming pool wastes, and ground water seepage. Other permits are for discharge of treated contaminated ground water, noncontact cooling water, and storm water. As of January 7, 2014, there were 314 dischargers covered under a construction storm water permit, and 1,235 dischargers covered under an industrial storm water permit.

A majority of the 117 NPDES permittees discharge directly into the Los Angeles River and a small number discharge into Burbank Western Channel, Compton Creek, and Rio Hondo. The largest numbers of general industrial storm water permits occur in the cities of Los Angeles (many within the community of Sun Valley), Vernon, South Gate, Long Beach, Compton, and Commerce. Metal plating, transit, trucking & warehousing, and wholesale trade are a large component of these businesses. This watershed has about twice the number of industrial storm water dischargers as does the San Gabriel River Watershed and the most in Los Angeles County and Ventura County watersheds.

At the time of this report there are a total of 314 construction sites enrolled under the construction storm water permit. The larger sites are located in the upper watershed, including the San Fernando Valley, and are fairly evenly divided between commercial and residential construction sites.

1.5.3 WATER QUALITY IMPAIRMENTS

The Clean Water Act (CWA) requires each State to assess the status of water quality in the State [Section 305(b)] and provide a list of impaired water bodies [Section 303(d)] to the U.S. Environmental Protection Agency

(USEPA) every two years. The majority of the Los Angeles River is considered impaired by a variety of point and nonpoint sources. The 2010 303(d) list implicates pH, ammonia, a number of metals, coliform bacteria, trash, odor, algae, oil, DDT, as well as other pesticides, and volatile organics for a total of 116 individual impairments (reach/constituent combinations). Some of these constituents are of concern throughout the length of the river while others are of concern only in certain reaches (Table 7).

Impairment may be a result of water column exceedances, excessive sediment levels of pollutants, or bioaccumulation of pollutants. The beneficial uses most often threatened or impaired by degraded water quality are aquatic life, recreation, groundwater recharge, and municipal water supply.

The CWA requires a Total Maximum Daily Load (TMDL) be developed to restore impaired water bodies to their full beneficial uses by allocating allowable loadings from point sources and nonpoint sources. TMDLs have been established for trash (2001), and bacteria (2012) for the Los Angeles River, for nitrogen compounds and related effects for the Los Angeles River (2004), for metals for the Los Angeles River and its tributaries (2006), and for nitrogen, phosphorus, trash, organochlorine pesticides, and PCBs for Los Angeles Area Lakes (2012).

EARLY WATER QUALITY MONITORING

Early water quality monitoring by the Los Angeles Department of Water and Power (LADWP) is reported following the creation of a water reclamation plant in 1929. Comparison of the recycled water from the plant to the untreated Los Angeles River water provided early water quality data for substances including ammonia, nitrogen, and suspended solids. Extensive testing and monitoring was done through the Headworks Groundwater Recharge Project to demonstrate that the recharge could meet regulatory requirements (Pinhey & Hogan, 2013).

TABLE OF CHEMICAL RESULTS WATER RECLAMATION PLANT

	Sewage Crude	Treated Sewage Settled	Stabilized	Fully treated	L. A. water untreated Aqueduct	Gallery	Per cent removal sewage to fully treat sewage compared with L.A. aqueduct
B. Coli	100,000	50,000	1	0.00	0.05	0.00	100+
Suspended Solids	418	77	3.2	0.0	6.0	0.0	100+
Turbidity	440	.....	3.3	0.2	25	0	100+
Oxygen demand	515	222	9.1	0.99	1.25	1.0	100+
Oxygen consumed	180	46	8.3	4.1	3.5	1.2	99.7
Organic nitrogen	48.1	37.3	5.08	2.25	1.3	0.5	99.8
Free ammonia	50	20	1.0	0.65	0.25	0.20	99.2
Ph	6.8	6.6	7.0	7.1	7.3	7.2	..

Reproduced from the Wastewater Professional.

1.5.4 THE LOS ANGELES RIVER WATERSHED MONITORING PROGRAM

Following the successful establishment of a watershed-wide monitoring program in the San Gabriel River Watershed in 2005, the LARWQCB required the Cities of Los Angeles and Burbank to develop a similar monitoring program for the Los Angeles River Watershed. The Los Angeles River Watershed Monitoring Program was developed in 2007-2008 by a work group consisting of representatives from stakeholder agencies throughout the watershed. The work group included representatives from the Cities of Los Angeles, Burbank and Downey, Los Angeles Water Board, Los Angeles County Department of Public Works, Friends of the Los Angeles River, Arroyo Seco Foundation, San Gabriel and Lower Los Angeles Rivers and Mountains Conservancy, U.S. Environmental Protection Agency (USEPA), Council for Watershed Health, Southern California Coastal Water Research Project (SCCWRP) and U.S. Forest Service. Following program development, the LARWQCB changed the City of Los Angeles and City of Burbank permit conditions to require implementation of the plan and the Council for Watershed Health was selected as program manager.

The monitoring program integrated as much as possible with existing monitoring, as well as including additional monitoring components: sampling at random sites throughout the watershed in order to assess overall watershed health; monitoring sites at high habitat value areas and at the base of subwatersheds; monitoring contaminants in the tissues of fish that are caught and consumed in lakes and streams; and monitoring fecal indicator bacteria at popular freshwater swimming sites. Each chapter of this State of the Watershed Report describes the monitoring plan and summarizes the results for each of these monitoring components, as well as providing recommendations for future monitoring.

Prior to the implementation of the Los Angeles River Watershed Monitoring Program, little was known about the condition of streams throughout the watershed. Permit conditions required discharges to monitor their effluents; however, these were located primarily in the mainstem and the lower watershed tributaries. These uncoordinated monitoring programs resulted in limited data comparability; lack of coordination on the constituents sampled, and unsynchronized data management and data quality. This State of the Los Angeles River Watershed report summarizes the results from the first five years of monitoring and provides a benchmark to assess the success of future management actions.

TABLE 7. Water quality impairments [303(d) list]

WATER BODY NAME	POLLUTANT
Los Angeles River Estuary (Queensway Bay)	Chlordane, DDT, and PCBs (sediment) Sediment Toxicity Trash
Los Angeles River Reach 1 (Estuary to Carson Street)	Ammonia Cadmium Coliform Bacteria Copper, Dissolved Cyanide Diazinon Lead Nutrients (Algae) Trash Zinc, Dissolved pH
Arroyo Seco Reach 1 (LA River to West Holly Ave.)	Benthic-Macroinvertebrates Coliform Bacteria Trash
Arroyo Seco Reach 2 (West Holly Ave to Devils Gate Dam)	Coliform Bacteria Trash
Compton Creek	Benthic-Macroinvertebrates Coliform Bacteria Copper Lead Trash pH
Echo Park Lake	Algae Ammonia Copper Eutrophication Lead Odor PCBs (tissue) Trash pH
Lincoln Park Lake (Carson to Figueroa Street)	Ammonia Eutrophication Lead Odor Organic Enrichment/ Low DO Trash

WATER BODY NAME	POLLUTANT
Los Angeles River Reach 2	Ammonia Coliform Bacteria Copper Lead Nutrients (Algae) Oil Trash
Rio Hondo Reach 1 (Confluence LA River to Santa Ana Fwy)	Coliform Bacteria Copper Lead Toxicity Trash Zinc pH
Rio Hondo Reach 2 (At Spreading Grounds)	Coliform Bacteria Cyanide
Aliso Canyon Wash	Copper Fecal Coliform Selenium
Bell Creek	Coliform Bacteria
Bull Creek	Indicator Bacteria
Burbank Western Channel	Copper Cyanide Indicator Bacteria Lead Selenium Trash
Dry Canyon Creek	Fecal Coliform Selenium, Total
Los Angeles River Reach 3 (Figueroa St. to Riverside Dr.)	Ammonia Copper Lead Nutrients (Algae) Trash
Los Angeles River Reach 4 (Sepulveda Dr. to Sepulveda Dam)	Ammonia Coliform Bacteria Copper Lead Nutrients (Algae) Trash

WATER BODY NAME	POLLUTANT
Los Angeles River Reach 5 (Within Sepulveda Basin)	Ammonia Copper Lead Nutrients (Algae) Oil Trash
Los Angeles River Reach 6 (Above Sepulveda Flood Control Basin)	Coliform Bacteria Selenium
McCoy Canyon Creek	Fecal Coliform Nitrate Nitrogen, Nitrate Selenium, Total
Tujunga Wash (LA River to Hansen Dam)	Ammonia Coliform Bacteria Copper Trash
Verdugo Wash Reach 1 (LA River to Verdugo Rd.)	Coliform Bacteria Copper Trash
Verdugo Wash Reach 2 (Above Verdugo Rd.)	Coliform Bacteria Trash
Legg Lake	Ammonia Copper Lead Odor Trash pH
Monrovia Canyon Creek	Lead
Peck Road Park Lake	Chlordane (tissue) DDT (tissue) Lead Odor Organic Enrichment/ Low DO Trash

REFERENCES

Bloom, P., et al. 2002. Avifauna along Portions of the Los Angeles River. FoLAR River Watch Biological Monitoring Program. 44 pgs.

DeCourten, F. 2010. Geology of Southern California. Textbook Supplement: Cengage Learning.

Garrett, K. 1993. The Biota of the Los Angeles River. Natural History Museum and California Department of Fish and Game. 120 pgs.

Hall, A., et al, 2012. Mid-Century Warming in the Los Angeles Region. Part 1 of the "Climate Change in the Los Angeles Region" project.

Los Angeles County Department of Public Works (LACDPW). 2006. Hydrology Manual.

Los Angeles County Department of Public Works (LACDPW). 1996. Los Angeles River Master Plan.

Los Angeles and San Gabriel Rivers Watershed Council (LASGRWC). 2001. Beneficial Uses of the Los Angeles and San Gabriel Rivers. Prepared by H. Trim. Edited by D. Green.

Los Angeles Regional Water Quality Control Board (LARWQCB). 1994. Water Quality Control Plan, Los Angeles Region. Los Angeles Regional Water Quality Control Board, Los Angeles, CA. http://www.swrcb.ca.gov/rwqcb4/water_issues/programs/basin_plan.

Pinhey, N., & Hogan, M. 2013. California's first indirect potable reuse treatment plant. Wastewater Professional. 49(2). 11-19.

Sheng, J., and Wilson, J.P. 2009. The Green Visions Plan for 21st Century Southern California. 22. Hydrology and Water Quality Modeling of the Los Angeles River Watershed. University of Southern California GIS Research Laboratory, Los Angeles, California.

United States Environmental Protection Agency (USEPA). 2012. Long Beach City Beaches and Los Angeles River Estuary Total Maximum Daily Loads for Indicator Bacteria. Approved March 2012.

Wohlgemuth, P.M. 2006. Hillslope Erosion and Small Watershed Sediment Yield Following a Wildfire on the San Dimas Experimental Forest, Southern California. Proceedings of the Eighth Federal Interagency Sedimentation Conference (8thFISC), April2-6, 2006, Reno, NV, USA.

CHAPTER 2

WHAT ARE THE CONDITIONS OF THE STREAMS IN THE LOS ANGELES RIVER WATERSHED?

2.1 BACKGROUND

The goal of this monitoring program is to assess the current condition of streams throughout the Los Angeles River Watershed for the purpose of informing past and future watershed management decisions. In the past century of development, the physical, biological, and chemical conditions of streams in the urban area of the Los Angeles River Watershed have been dramatically altered. In contrast, streams in the more remote areas of the upper watershed maintain some pre-urbanization integrity, providing an opportunity to assess a gradient of conditions across the watershed. We provide a summary and assessment of the current condition of streams over the five years from 2008 through 2012 as a comprehensive baseline for assessing future management actions.

2.2 MONITORING RESULTS (2008-12)

To determine the condition of streams in the Los Angeles River Watershed, a total of fifty sites, ten sites annually, were sampled from 2008 through 2012 (Figure 11). These sites were randomly selected and weighted to ensure that the three watershed subregions were adequately represented: the natural portions of the upper watershed, the effluent-dominated reaches of the mainstem, and the urban tributaries of the lower watershed.

The scope of this monitoring is confined to assessing the condition of perennial streams; a perennial stream is one that has continuous flow in parts of its streambed all year round. The monitoring program design is consistent with regional and statewide Perennial Streams Assessment (PSA) programs that are built upon earlier programs,

FIGURE 11. IBI scores at random sites sampled 2008 through 2012

namely US EPA's Environmental Monitoring and Assessment Program (EMAP) and California's Monitoring and Assessment Program (CMAP). Samples are collected during the dry weather only (May through July) to increase the likelihood that streams are truly perennial.

Bioassessment is the key component of these monitoring programs, using resident aquatic biota as indicators of the biological integrity of streams. Bioassessment is combined with chemical, toxicological, and physical habitat characteristics of the sites to provide a multiple lines of evidence (MLOE) approach for assessing stream condition.

2.2.1 BIOLOGICAL CONDITION

The biological condition of streams was assessed using the Southern California Index of Biological Integrity (IBI). This index was developed to determine the response of in-stream biological communities, in this case aquatic invertebrates, to physical and chemical stressors. During the 5-year monitoring period, benthic macroinvertebrate communities in the more natural streams in the upper watershed had the highest IBI scores *(Figure 12)* and were most similar to those at "undisturbed" reference sites throughout Southern California.

The benthic macroinvertebrate communities in the upper watershed exhibited a wide range of feeding strategies and were characterized by organisms that were pollution sensitive. In contrast, the biological communities in the lower urban and effluent-dominated reaches were more degraded, as evidenced by lower IBI scores, fewer feeding strategies, and the dominance of organisms that were more tolerant of pollution.

FIGURE 12. Biological condition (IBI scores) for different watershed subregions

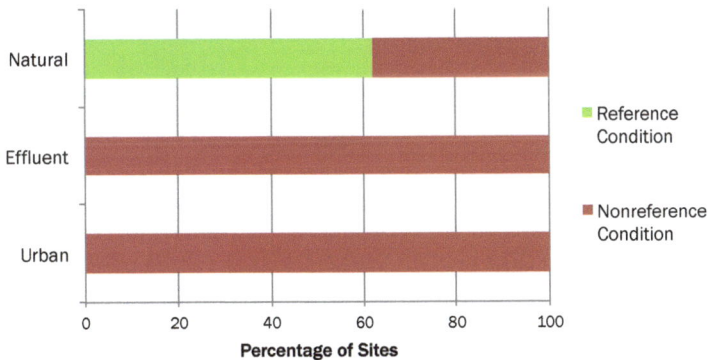

ASSESSING BIOLOGICAL CONDITION IN CALIFORNIA'S STREAMS

From 2008 thru 2012 the Southern California Index of Biological Integrity (IBI) was used to measure stream condition using in-stream biological communities. The IBI is a multi-metric index that incorporates seven biological metrics that respond to different environmental stressors. The index was developed using data collected from over 250 sites throughout southern California, including both relatively pristine reference sites and sites influenced by human activities (Ode et. al. 2005).

The IBI rates sites on a scale of 0 to 100, with scores over 39 indicating a stream condition where the biological community is similar in complexity to reference sites in the region. The minimally disturbed reference sites used in its development, however, were concentrated in higher elevation and higher gradient locations. Therefore, the index may not be as useful for interpreting biological integrity in the lower portion of the watershed.

Most recently, the California Stream Condition Index (CSCI) was developed to address this, and other deficiencies in the IBI. The CSCI was created from a larger and broader dataset to establish site-specific expectations, as opposed to region-wide, and is therefore applicable throughout the state. The index has two components: a measure of the number of species observed at a site compared to what is expected (O/E), and a measure of the community structure using similar metrics to the IBI. LARWMP will use CSCI to assess the biological community condition beginning in the 2013 monitoring year.

2.2.2 AQUATIC CHEMISTRY AND TOXICITY

A comparison of water quality parameters among the three watershed subregions from 2008 thru 2012 demonstrates the spatial differences in water quality (Table 9).

- There were few exceedances of dry-weather Basin Plan standards over the 5-year period. Exceptions included elevated pH values in the channelized portions of the lower watershed, copper and selenium in the urban tributaries, and impaired biological communities throughout the watershed.

- Much of lower Los Angeles River Watershed has been channelized and has little riparian vegetation to provide shade, so water temperatures can reach as high as 36°C (97°F) during summer. Direct sunlight on the low gradient, slow-moving, nutrient-rich water leads to accelerated algal growth and high rates of photosynthesis. The high rate of photosynthesis during daylight hours produces oxygen and consumes carbon dioxide, causing the elevated pH and DO observed during daylight hours.

- Nutrients and metals were consistently lower at sites in the upper watershed compared to the lower tributaries and the mainstem. Although nitrogenous compounds were higher in the effluent dominated and urban reaches, they did not exceed water quality standards. Organophosphorus and pyrethroid pesticides were nearly always below method detection limits (i.e., non-detect).

- A total of 41 water samples were tested for acute and chronic toxicity using water fleas (Ceriodaphnia dubia). The greatest number of chronic toxic endpoints was observed at natural sites in the upper watershed-where 12 of the 19 samples (63%) showed reproductive toxicity. Reproductive toxicity was also observed in 50% of samples from the lower watershed tributaries. All of the acute (survival) toxic endpoints were in the lower tributaries and upper watershed, with no acute toxicity measured in the effluent-dominated reaches.

2.2.3 INDICATOR BACTERIA

Bacteria levels in urbanized watersheds can be highly erratic and dependent on a multitude of potential human and non-human sources (CREST, 2010). LARWMP sampled seven sentinel sites throughout the lower watershed from 2009 through 2012 to determine the degree to which water quality objectives are being met or exceeded, as well as reach-specific information relevant to the bacteria TMDL in the watershed (CREST, 2010). Where possible, pre-LARWMP monitoring locations were retained and new sites established at the bottom of each mainstem reach to provide the ability to track progress toward reach-specific TMDL targets.

In 2009, LARWMP commenced twice-weekly year-round monitoring for fecal coliforms, total coliforms, and Enterococcus (AB411 indicators) in the Los Angeles River estuary. These data will provide a basis for evaluating the potential relationship between the river discharge, offshore plumes, and beach contamination along the coastline.

With the exception of the estuary, the aforementioned sites are not designated REC-1 recreational swim sites and public access is not allowed. Bacteria concentrations measured at these sites, however, are compared against REC-1 standards to provide context.

LARWMP results to date show that fecal indicator bacteria are ubiquitous throughout the watershed and exceedances of the single sample REC-1 standard (Table 8) is common at all sites (Figure 13). The greatest frequency of exceedances occurring in the highly urbanized Tujunga Wash, Burbank Channel, and Cerritos Channel areas. The lowest bacteria concentrations, and fewest exceedances, occurred at sites at or below POTW discharges.

TABLE 8. Indicator bacteria REC1 standards for freshwaters

Indicator	Single Sample Standard	30-Day Geometric Mean
Total Coliform	10,000	1,000
E. coli	235	126
Enterococcus bacteria	104	35

The 30-day geometric mean provides an indication of how persistent elevated bacterial concentrations are at a site. Sentinel sites typically exceeded the 30-day geometric mean REC1 standard during each month and these findings are consistent with those reported by CREST (2010).

TABLE 9. Water quality within the different watershed subregions for the period 2005-2009. Yellow indicates more than one exceeedance.

PARAMETER (UNITS)	EFFLUENT (n=9)			URBAN (n=19)			NATURAL (n=21)		
	Mean	Range	# Exc.	Mean	Range	# Exc.	Mean	Range	# Exc.
General Chemistry									
Dissolved Oxygen (mg/L)	11.2	4.89 - 17.45	1	10.2	7.25 - 16.81	0	8.3	6.66 - 10.48	0
pH	8.8	7.96 - 9.38	7	8.7	7.42 - 10.8	7	8.1	7.1 - 8.51	0
Specific Conductivity (uS/cm)	1130	962 - 1355		1369	7.75 - 3681		402	245 - 699	
Temperature (°C)	23.5	13.36 - 32.8		24.2	13.84 - 36.14		16.6	10.97 - 25.03	
Mean Slope (%)	0.4	0.12 - 0.908		1.1	0.114 - 4		3.3	0.85 - 8.868	
Discharge (m³/sec)	2.1	0.15 - 4.496		0.7	0.001 - 5.546		1.0	0.002 - 11.498	
Alkalinity as CaCO₃ (mg/L)	148	100 - 206		402	74 - 4520		189	119 - 270	
Total Suspended Solids (mg/L)	30	7.74 - 93.6		77	4.5 - 632		4	1.2 - 8.4	
Dissolved Organic Carbon (mg/L)	6.6	6.2 - 7.39		12.4	1.79 - 33.4		2.9	1.38 - 6.83	
Total Organic Carbon (mg/L)	10.9	6.85 - 32.4		13.0	2.64 - 38		11.6	1.48 - 102.22	
Nutrients (mg/L)									
Ammonia	N.D		0	N.D		0	N.D		0
Nitrate as N	3.12	0.98 - 5.2	0	1.51	0.07 - 4.26	0	0.18	0.04 - 0.53	0
Total Kjeldahl Nitrogen	2.16	1.6 - 2.8		2.22	0.14 - 5.8		0.36	0 - 1.73	
Orthophosphate as P	0.24	0.09 - 0.62		0.24	0.05 - 1.52		0.06	0.025 - 0.12	
Dissolved Metals (ug/L)*									
Arsenic	1.9	1.13 - 2.8	0	3.1	0.405 - 13.02	0	1.4	0.11 - 4.44	0
Chromium	15.2	1.56 - 121	0	43.2	0.22 - 215	0	2.9	0.21 - 7.26	0
Copper	7.7	2.125 - 15.1	1[b]	9.6	0.8 - 26	2[b]	1.7	0.275 - 3.65	0
Iron	43.2	14.56 - 93	0	54.4	2.5 - 195	0	51.5	0.67 - 337	0
Lead	0.55	0.055 - 1.47	0	1.6	0.055 - 21.63	1[b]	0.11	0.05 - 0.21	0
Mercury	0.0019	0.0018 - 0.002		0.009782	0.0018 - 0.0468		0.005033	0.0018 - 0.0412	
Nickel	5.7	2.53 - 7.81	0	8.1	0.65 - 50	0	1.5	0.61 - 3.87	0
Selenium	1.8	0.4 - 3.5	0	1.9	0.1 - 10.585	2[b]	0.13	0.1 - 0.25	0
Zinc	36.6	20.7 - 54.7	0	27.4	1.47 - 293.83	0	3.3	0.73 - 15.027	0
Biological Condition									
IBI	10	4.29 - 23	9	12	0 - 28.6	19	39	20.02 - 77.22	13
Riparian Habitat Condition									
CRAM	37	27 - 47		37	27 - 64		74	55 - 99	
Toxicity									
Toxic Endpoints (Acute/Chronic)		n=6 0/1			n=16 1/8			n=19 1/12	

WATER QUALITY OBJECTIVES
Dissolved Oxygen: 5 or ≥ 7
pH: 6.5 – 8.5
Nitrate as N: 8 mg/L
Exc.: Number of exceedances

* Hardness Adjusted Dissolved Metals compared to the California Toxic Rule (CTR)
a CTR Acute Threshold Value
b CTR Chronic Threshold Value

FIGURE 13. *E. coli* sampling locations

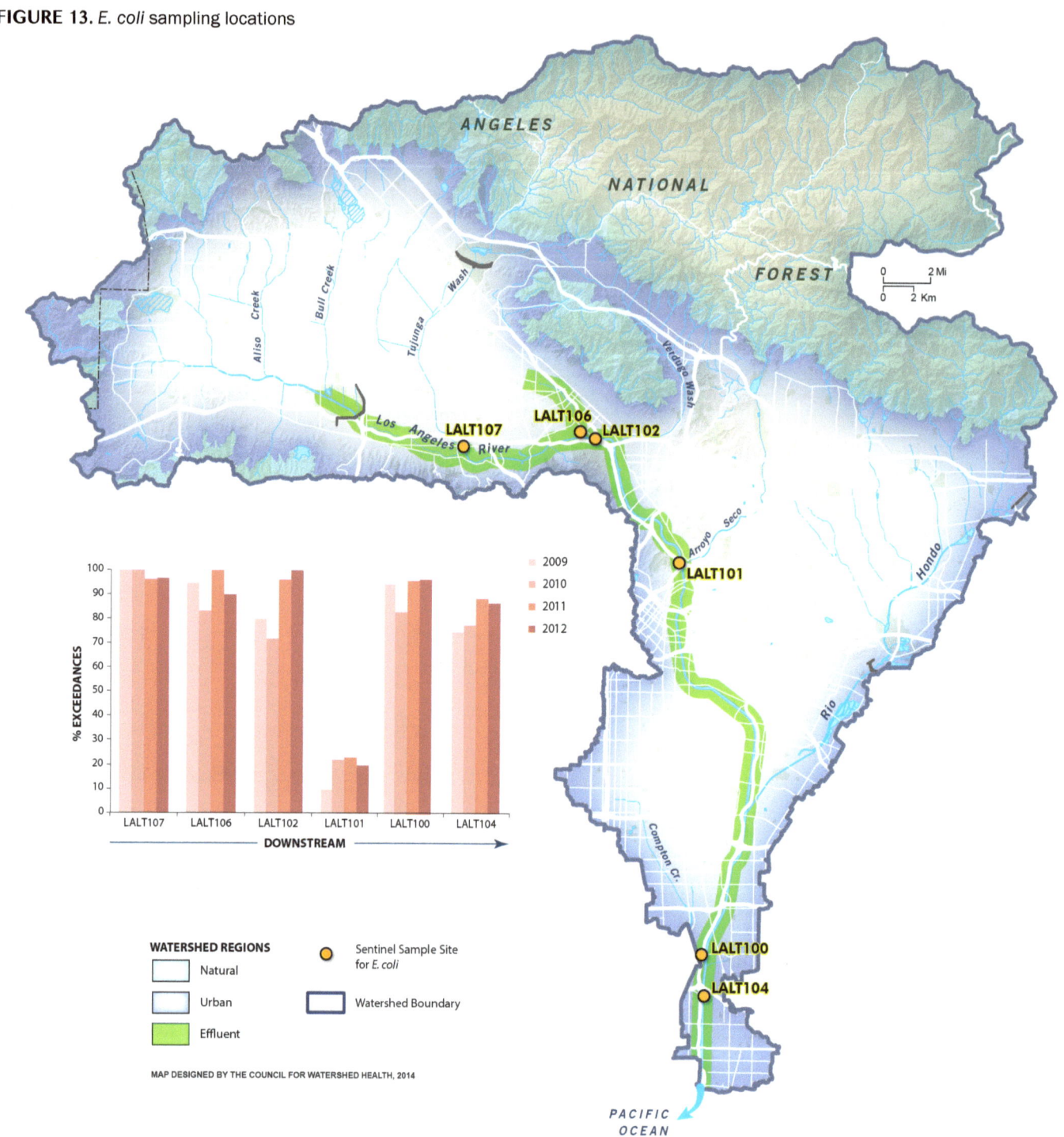

BACTERIA IN THE LOS ANGELES RIVER ESTUARY

The estuary is heavily impacted by trash and other pollutants flowing down the Los Angeles River. Large booms have been installed to collect trash before flows from the watershed enter the estuary. The Los Angeles River can contribute significant amounts of bacteria to the estuary and ultimately, Long Beach City beaches. Other than river flows, sources of bacteria to the estuary include wildlife, predominantly birds and waterfowl, and Municipal Separate Storm Sewer System (MS4) dischargers. Although the estuary has not been identified as impaired by the LARWQCB, it has been confirmed as impaired through data analyses and is included in the above-mentioned TMDL as an unaddressed source of bacteria that has the potential to impact the Long Beach City beaches.

LARWMP began twice-weekly year-round monitoring for fecal coliforms, total coliforms, and *Enterococcus* (the AB411 indicators) in the Los Angeles River estuary in 2009. Total Coliform bacteria are persistent throughout the dry-weather monitoring period and always exceed water quality standards *(Table 10)*. *Enterococcus* and *E.coli* bacteria are also persistent at elevated concentrations with few exceptions in 2010 (ie., August and September).

2.2.4 STORMWATER QUALITY

During the wet season, large volumes of stormwater enter the Los Angeles River and carry pollutants from diffuse sources through the system to the Pacific Ocean. The LARWQCB regulates this stormwater under an MS4 permit. The 2001 permit was issue to the LACFCD, the County of Los Angeles and 84 cities therein (except the City of Long Beach, which was issued a separate permit). In its role as a Principal Permittee, the LACFCD was required to coordinate and facilitate stormwater management activities such as stormwater quality monitoring. The results from this monitoring are summarized in this report. The most recent MS4 permit, effective November 28, 2012, relieves the LACFCD from

TABLE 10. 30-day geometric mean bacteria concentrations (MPN/100 mL) at the Los Angeles River estuary. See *Table 8* for exceedance thresholds.

	May	Jun	Jul	Aug	Sep
E. coli					
2009	391	2732	1222	1365	1811
2010	51	129	70	70	99
2011	4817	276	161	47	85
2012	52	127	139	93	981
Enterococcus					
2009	65	3194	39	37	43
2010	45	33	22	28	69
2011	3088	47	55	54	27
2012	37	44	79	32	171
Total Coliform					
2009	11471	16889	19364	17240	21374
2010	3617	7617	10978	15794	13548
2011	11689	7100	14353	10108	14778
2012	2858	13272	13369	18885	24000

CLEANER RIVERS THROUGH EFFECTIVE STAKEHOLDER-LED TMDLS (CREST)

Prior to the LARWMP, much of the Los Angeles River and its tributaries were included on the state/federal list of impaired water bodies, the 303(d) list, for fecal coliform bacteria. The Clean Water Act specifies that water bodies on the 303 (d) list are required to develop Total Maximum Daily Loads (TMDLs) with the goal of achieving water quality objectives. A TMDL for indicator bacteria was developed by the Los Angeles Regional Water Quality Control Board (RWQCB) in cooperation with the Cleaner Rivers through Effective Stakeholder-led TMDLs (CREST) stakeholder group. A comprehensive Bacteria Source Identification (BSI) study was undertaken for the purpose of the TMDL and highlighted the following sources and contributions (CREST 2010).

- Approximately 85% of storm drain samples exceeded the 235 MPN/100 mL objectives for *E.coli*.

- Despite the fact that storm drains and tributaries contribute roughly 13% of thé flow in the Los Angeles River, and WRPs contribute approximately 72%, discharges from storm drains contribute almost 90% of the *E. coli* loading from point sources to the river during dry weather.

- MS4 discharges are the principal source of bacteria to the Los Angeles River and its tributaries in both dry weather and wet weather.

- Although hundreds of storm drain outfalls discharge varying levels of bacteria to the LA River during dry weather, other in-channel sources including birds, homeless persons, and perhaps environmental re-growth also are significant.

TABLE 11. Summary of water quality during storms: number of samples (% exceedance).

	Aliso Creek	Bull Creek	Burbank W. Channel	Verdugo Wash	Arroyo Seco Channel	Rio Hondo Channel
	n (% exceedance)					
TSS (mg/L)	9 (0)	9 (0)	9 (0)	9 (0)	9 (0)	9 (0)
Fecal coliform (MPN/ 100 mL)	9 (100)	9 (100)	9 (100)	9 (55)	9 (100)	9 (55)
Nitrate/ Nitrite (mg/L)	9 (0)	9 (0)	9 (11)	8 (0)	8 (0)	9 (0)
Ammonia (mg/ L)	8 (0)	9 (0)	8 (0)	8 (0)	8 (0)	8 (0)
Phosphorus, Total (mg/L)	8 (0)	9 (0)	8 (0)	9 (0)	8 (0)	8 (0)
Cadmium, Total (Ug/L)	8 (0)	9 (0)	9 (0)	9 (0)	8 (0)	9 (10)
Copper, Total (Ug/L)	8 (75)	9 (44)	9 (88)	9 (88)	8 (75)	9 (77)
Lead, Total (Ug/L)	8 (13)	9 (22)	9 (33)	9 (55)	3 (38)	9 (22)
Zinc, Total (Ug/L)	8 (50)	9 (33)	9 (55)	9 (33)	8 (50)	9 (44)
Diazinon (Ug/L)	9 (60)	9 (33)	9 (33)	8 (13)	9 (44)	9 (33)

Source: LACDPW Annual Monitoring Reports 2002-2003 and 2003-2004

this role and regulates individual cities, as well as the County operating for unincorporated areas. The permit encourages municipalities to join together in watershed management groups and provides the flexibility to customize and implement stormwater monitoring and management programs.

As a requirement of the permit, LACFCD collected water quality data from subwatersheds in the Los Angeles River Watershed during 2002-2003 and 2003-2004. The following six tributaries were monitored: Aliso Creek, Bull Creek, Burbank Western Channel, Verdugo Wash, Arroyo Seco Channel, and Rio Hondo Channel. *Table 11* summarizes two years of results of wet-weather monitoring at these locations for common constituents of concern in stormwater. Fecal coliform bacteria were ubiquitous across all of the monitoring stations and typically always exceeded the Basin Plan water quality standard. Total copper, total zinc, and diazinon also regularly exceeded regulatory standards. A more comprehensive summary of the monitoring results is provided in annual stormwater monitoring reports *(http:// dpw.lacounty.gov/wmd/NPDES/report_directory.cfm).*

2.2 SUMMARY AND NEXT STEPS

Over the five years of monitoring, we found a strong positive relationship between the condition of the biological communities and physical habitat conditions in the Los Angeles River watershed. While nutrients and metals were elevated in the lower tributaries

and mainstem, they did not exceed water quality objectives. Despite this, there are potentially harmful compounds that were not monitored in this study that may still impact biological communities such as benthic macroinvertebrates and organisms that prey on them. Future monitoring will focus on more discernibly linking the condition of the biological communities to the physical, chemical, and toxicological stressors through tools and techniques that identify the specific stressors to which the biological communities have been exposed.

Prior to 2011, sites were not distributed to represent the number of stream miles represented by each subregion and were evenly distributed across the three defined subregions. Beginning 2011, the program reallocated the number of randomly selected sites within these subregions so that five sites will be sampled in the lower watershed (urban tributaries), four sites in the upper watershed (natural) and one site on the effluent-dominated reaches annually.

Over the next five years the LARWMP will continue to support and coordinate with larger regional monitoring efforts such as the Southern California Stormwater Monitoring Coalition (SMC) program and the State Water Board's Surface Water Ambient Monitoring Program (SWAMP). Over the next two years, the SMC program will be revising its monitoring plan and assessment tools to reflect regional priorities and the development of new assessment tools. The LARWMP will adjust relevant monitoring program components accordingly. In future years, for example, LARWMP will assess stream condition using the California Stream Condition Index (CSCI) and

Post-Station Fire monitoring in the Angeles National Forest near Alder Creek.

may continue to monitor attached algae at random sites. Although the target of this monitoring is perennial streams, there is no evidence that all of the streams monitored in the Los Angeles River watershed fulfill the definition of perenniality and future monitoring will focus on identifying these streams as an additional indicator of biological condition.

LARWMP will continue to support efforts to determine the effects of post-fire runoff on surface water quality in the watershed. The Station Fire, which started on August 26, 2009, was the largest fire in the recorded history of Angeles National Forest (est. 1892) and the tenth largest fire in California since 1933. The fire ignited near the U.S. Forest Service ranger station on the Angeles Crest Highway (State Highway 2) and burned 160,577 acres (251 sq. mi). LARWMP sampled three sites in the burn area prior to the fire and has continued to monitor these sites annually since the fire.

In 2011, LARWMP commenced trash assessments at targeted and random sites throughout the watershed to assist the SMC region-wide trash assessment study. The monitoring data assisted the Los Angeles City Council in the decision to ban plastic grocery bags in the City of Los Angeles. LARWMP is also working with researchers at local and national universities to increase our understanding of the economic, scientific, and health implications of the impending requirement to use molecular monitoring technologies for the quantification of bacteria at freshwater swimming sites.

REFERENCES

Ode, R.E., Rehn, A.C., and May. J.T. 2005. A Quantitative Tool for Assessing the Integrity of Southern Coastal California Streams. Env. Man., Vol. 35, No. 4, pp. 493–504.

Cleaner Rivers through Effective Stakeholder TMDLs (CREST). 2010. Draft Los Angeles Watershed Bacteria TMDL – Technical Report Section 5: Dry Weather Linkage Analysis. Available at: http://www.crestmdl.org/reports/pdf/DRAFT-LINKAGE-LAR-Bact-TMDL-Technical-Report-041310distributed.pdf

FIGURE 14. Estuary, targeted and high-value habitat sites

CHAPTER 3

ARE CONDITIONS AT LOCATIONS OF UNIQUE INTEREST GETTING BETTER OR WORSE?

Compton Creek confluence/Los Angeles River confluence (facing southwest).

3.1 BACKGROUND

LARWMP assessed how habitat conditions might be changing over time for habitats of unique interest. The fourteen designated sites are assessed annually: four confluences representing the major subwatersheds, the Los Angeles River estuary, and nine wetland habitats that represent unique areas of special concern in the watershed. Aquatic chemistry, toxicity, biota, and physical habitat data were collected annually from 2009 to 2012 (Figure 14).

Four of the targeted watershed sites were established upstream of confluence points in the upper and lower watershed to provide information regarding water quality trends over time. These four sites differ from the random sites in the previous chapter because their locations are fixed and are sampled each year. These data are being used to assess temporal trends and whether changes in these trends can be attributed to natural, anthropogenic, or watershed management changes.

Prior to the initiation of this program, there had been no coordinated sampling effort in the estuary. Samples are collected at the mouth of the Los Angeles River estuary near Queensway Bridge. The program was designed so that data assessment tools specific to the State's sediment quality objectives (SQOs) (SCCWRP 2009) could be used to determine the condition of habitat in the estuary.

Nine sites were identified by workgroup members to represent unique habitats of high value for their relatively

natural habitats in otherwise heavily urbanized areas. They provide a measure of natural background or provide context against which trends in other portions of the watershed can be evaluated. The primary goal of this component of the program is to track trends over time and provide early warning of potential degradation so that management action can be taken.

3.2 CONFLUENCE SITES

There were no discernible temporal trends in water chemistry at the four confluence sampling locations during the initial monitoring period. Spatially, however, both total and dissolved organic carbon were greatest at the Western Burbank Channel (LALT503) and Compton Creek (LALT502) confluences. Arsenic, copper, nickel, and mercury were routinely greater at the Western Burbank Channel (LALT503), while selenium was greatest at the Rio Hondo confluence (LALT500).

We evaluated toxicity at target sites using the water flea (Ceriodaphnia dubia, 7-day chronic test) toxicity test. Acute (survival) toxicity was not detected in samples collected from 2009 through 2012. Chronic (reproductive) toxicity was measured at site LALT501 at the confluence of the Arroyo Seco with the Los Angeles River in 2009, 2011, and 2012, and at the Western Burbank Channel (LALT503) on one occasion in 2009 (Figure 15). Toxicity at the Arroyo Seco confluence appears to be persistent and future monitoring will focus on identifying the source of this toxicity.

THE CALIFORNIA RAPID ASSESSMENT METHOD (CRAM)

The California Rapid Assessment Method (CRAM) was developed to provide biologists and ecologists a quick way to evaluate the complex ecological condition of wetlands and riverine systems using a finite set of observable field indicators, such as plant community composition and structure, hydrology, physical structure, and buffers (Stein, et al. 2009). CRAM assesses and scores wetland condition with respect to four overarching attributes: Buffer/ Landscape Context, Hydrology, Physical Structure, and Biotic Structure. *http://www.cramwetlands.org*

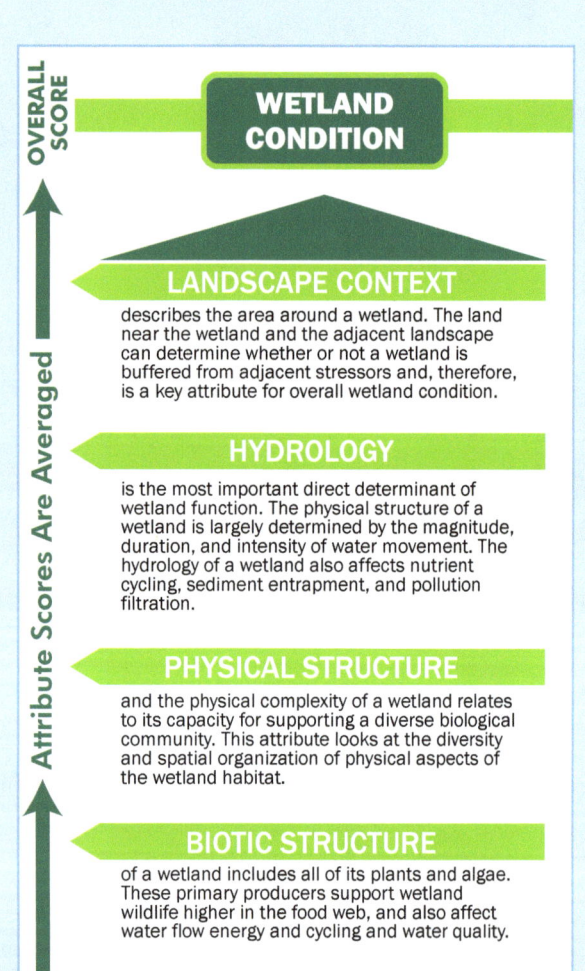

FIGURE 15. Chronic (reproductive) toxicity at target sites.

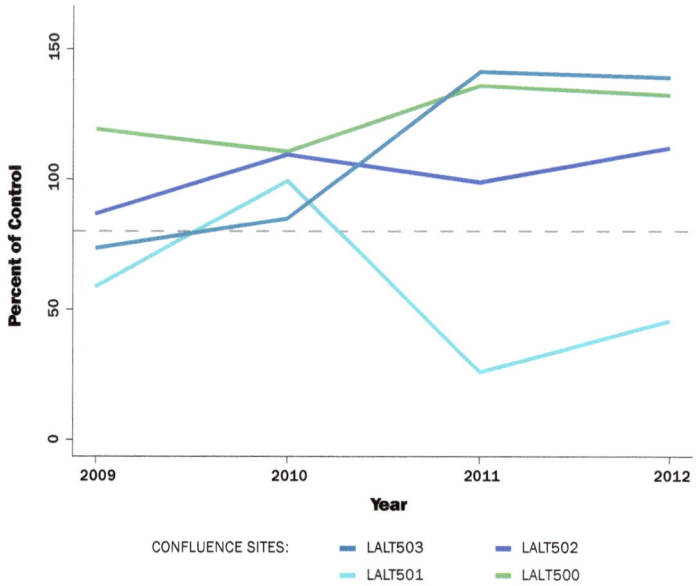

FIGURE 16. Southern California IBI scores at target sites. The dashed grey horizontal line (39) represents altered biological communities.

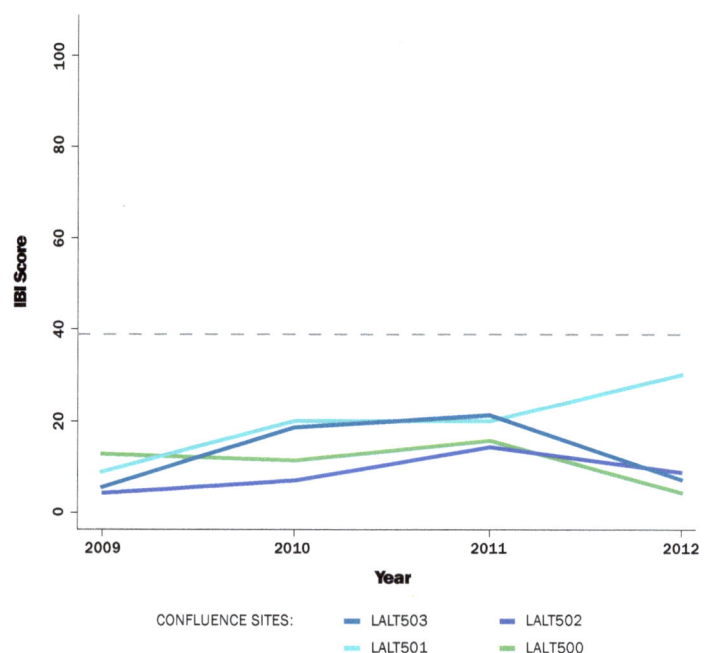

TABLE 12. Category scores for each indicator and MLOE site evaluation for estuary site EST2.

	2009	2010	2011	2012
Chemical Exposure	High Exposure	Moderate Exposure	Moderate Exposure	Moderate Exposure
Benthic Disturbance	Moderate Disturbance	Reference	Low Disturbance	High Disturbance
Toxicity	Moderate Toxicity	Moderate Toxicity	Nontoxic	Nontoxic
SITE ASSESSMENT	Clearly Impacted	Likely Impacted	Unimpacted	Possibly Impacted

IBI scores at the confluence sites scored in the 'poor' and 'very poor' range for all years compared to 'reference site' conditions in southern California. These sites are located in concrete-lined channels in the urbanized subregion. In addition to good water quality conditions, healthy benthic macroinvertebrate communities require complex in-stream and riparian cover and a wide and undisturbed riparian and buffer zone. CRAM scores followed the pattern in IBI scores, highlighting connection between the condition of the biological community and the quality of physical habitat *(Figure 16)*. CRAM scores reflected degraded riparian habitat at all sites. Compton Creek (LALT502) is the only site with an unlined (no cement) bottom which provides habitat for plants, invertebrates, and wildlife and this is reflected in the higher CRAM score compared to the other sites.

3.3   LOS ANGELES RIVER ESTUARY

The State of California Sediment Quality Objectives (SQOs) is a Multiple Lines of Evidence (MLOE) approach for assessing the exposure of organisms to sediment contamination. Three assessments provide evidence that a site is impacted: sediment chemistry, sediment toxicity, and the condition of resident infauna species. Assessment results are integrated into a rank score providing insight into the whether an embayment habitat is similar to reference conditions or has been degraded as a result of sediment contamination (SCCWRP, 2009).

The integrated scores for sediment chemical exposure, toxicity, and benthic disturbance were calculated at the estuary from 2009- 2012 *(Table 12)*. Sediment quality was highly variable across the four-year period ranging from 'clearly impacted' to 'unimpacted'. The estuary is located at the most downstream portion of the watershed and receives contaminants associated with urban runoff and point source and nonpoint source discharges from the watershed. These contaminants are received mainly during wet-weather storms from November through March.

Benthic disturbance was the most variable indicator ranging from 'highly disturbed' in 2012 to 'reference' in 2010. These results reflect the fact that this is a highly modified estuary with little protection from scoring after storm events and daily tidal flushing. These natural events transport contaminants both to and from the estuary, constantly changing the habitat conditions for benthic infauna. As a result, this habitat is highly unstable and in a constant state of flux.

3.4   HIGH-VALUE SITES OF UNIQUE CONCERN: RIVERINE AND ESTUARINE WETLANDS

Wetland habitats are particularly important for their relatively natural state in otherwise heavily urbanized areas of the watershed. The Program assesses condition of these nine sites over time to provide information that can result in the development of either restorative or protective management decisions. Specific areas of value and/or at-risk habitat established for the Program are: Sepulveda Basin, Glendale Narrows, the upper portion of Tujunga Wash, the upper portion of Arroyo Seco, Eaton Wash, and Golden Shores Wetland at the estuary. Minimally impacted sites that provide a measure of natural background or context are: USGS gaging station in Arroyo Seco, Alder Creek, and Haines Creek Pools and Stream that are above campgrounds and other human influences. These sites are relatively accessible by road and have late-season flows. A description of each site is provided below. Initially, CRAM assessments were conducted annually *(Figure 17)*. More recently, conditions are being assessed on a three-year cycle acknowledging the inherently slow rate of change in the component indicators.

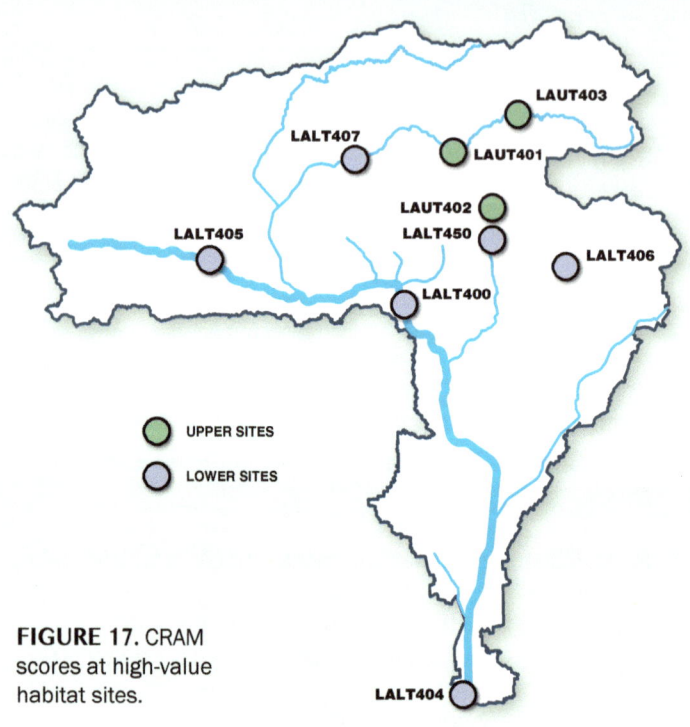

FIGURE 17. CRAM scores at high-value habitat sites.

UPPER SITES

LOWER SITES

UPPER HIGH-VALUE SITES

| | 0 | 20 | 40 | 60 | 80 | 100 |

LAUT401 (2009)
(2010)
(2011)

AVG.=73

LAUT402 (2009)
(2010)
(2011)

LAUT403 (2009)
(2010)
(2011)

LOWER HIGH-VALUE SITES

| | 0 | 20 | 40 | 60 | 80 | 100 |

LALT400 (2009)
(2010)

AVG.=62

LALT404 (2009)
(2010)

LALT405 (2009)
(2010)
(2011)

LALT406 (2009)
(2010)

LALT407 (2009)
(2010)

LALT450 (2009)
(2010)

TUJUNGA SENSITIVE HABITAT (LALT401)

The Tujunga Sensitive Habitat site is located downstream of the Big Tujunga Dam in a relatively undisturbed, upper watershed riparian zone. Big Tujunga Canyon is high in species richness, including 38 recorded threatened and endangered species of amphibians, reptiles, fish, and birds and twenty-four plants. This area burned during the 2009 Station Fire and therefore provides an opportunity to assess the post-fire recovery process for the riparian corridor and the surrounding buffer zones. Since this site is difficult to access it is not heavily-used by the public for recreation.

CRAM scores at this site have steadily increased from 2009 (64) through 2012 (77) owing to improved biotic structure of the plant communities, recovery of the physical structure following the heavy scouring that occurred from post-fire winter storms in 2010, and continued improvement in the hydrology of the site. The buffers surrounding the site have been in excellent condition throughout the period.

UPPER ARROYO SECO (LALT402)

The Arroyo Seco Watershed begins at Red Box Saddle in the Angeles National Forest near Mount Wilson in the San Gabriel Mountains. Much of the watershed contains nearly pristine habitat area and as a result, the hiking trails running along its length are very popular with the public for hiking and cycling. The biological condition score at this site (as measured by the SoCA IBI) is one of the highest in southern California. This site was devastated by the 2009 Station Fire and has been the location for an ongoing post-fire recovery study.

The CRAM score at this site have ranged from 50 to 87, with an average of 74 from 2009 to 2011. We have seen a slight improvement since the Station Fire. The consequences of the fire, which modified the physical structure of the streambed and surrounding riparian zone, included loss of complexity and erosion. The buffer zones and hydrology of the site changed only slightly during the same period.

ALDER CREEK (LAUT403)

Alder Creek is located in the upper reaches of the Los Angles Watershed and is the highest elevation of the unique habitat sites. Due to the remoteness, it provides a sentinel for conditions in the relatively undisturbed upper watershed.

CRAM scores at this site were stable over the period from 2009 to 2012 (average 75). The buffer zones at this upper watershed site were in excellent condition with essentially no human influence. The site is a wash and therefore the physical structure of the streambed and banks scored lower than the other attributes. The biotic structure scores were also lower owing to a lack of vegetative layering and sparse vegetative density.

GLENDALE NARROWS (LALT400)

The Glendale Narrows is an approximately eight-mile long section of the Los Angeles River adjacent to Griffith Park, Los Feliz, Atwater Village, and Elysian Valley. It is earthen bottom as a result of the high water table; however, the banks are shored with concrete. The earthen bottom provides a complex streambed composed of cobble, boulders, and sand that support diverse plant, bird, and fish communities. Frontage roads along both banks serve as walking and cycling paths for the public. Recently, this section of the river was opened to recreational boating.

CRAM scores at this site averaged 53 during 2009 and 2010 owing to its poor physical structure, lack of buffer zones, and highly modified hydrology. The plant community is dominated by invasive species.

THE GOLDEN SHORES WETLAND (LALT404)

Golden Shores Wetland was constructed in 1997 as part of mitigation for wetlands that were destroyed in Long Beach Harbor. The 6.4-acre wetland at the mouth of the Los Angeles River includes both intertidal and subtidal habitat and is one of the few tidally influenced wetlands in southern California. These habitats are important to the coastal ecosystem because they serve as highly productive habitats for fish, waterfowl, and plants. The entire perimeter of the wetland is protected by rip-rap levees with a single southern inlet connected to the Los Angeles River estuary. The buffer zone surrounding the wetland includes parking lots and port infrastructure.

The overall CRAM score (61) for this wetland was below the average score (75) for estuarine wetlands that were surveyed for a state-wide calibration study conducted by SCCWRP. The habitat complexity typical of a perennial wetland has been lost; the limited freshwater flow into the wetland consists mostly of urban runoff including runoff from the port. Conversely, the biotic structure of the wetland is relatively good, with several native plants species present.

SEPULVEDA BASIN (LALT405)

Sepulveda Basin upstream of Sepulveda Dam is mostly operated under lease by the City of Los Angeles Department of Recreation and Parks. The 225-acre area includes sports fields, agriculture, golf courses, a fishing lake, parklands, a water reclamation treatment facility, and a wildlife refuge. The 3-mile reach of river upstream of the dam is unlined with relatively natural riparian zones.

CRAM scores at this site averaged 60 from 2009 to 2011. The buffer zones on this reach of the Los Angeles River are wide and relatively undisturbed; however, the physical structure of the stream bottom and banks are highly modified and lack complexity. The water source for this reach is nearly 100% tertiary-treated effluent from the D.C. Tillman Water Reclamation Plant during the dry season.

EATON WASH (LALT406)

Eaton Canyon begins at the Eaton Saddle near Mount Markham and San Gabriel Peak in the San Gabriel Mountains. Its drainage flows into the Rio Hondo to the Los Angeles River. The Eaton Canyon Natural Area Park covers 190-acres where Eaton Creek forms a 50-foot waterfall from the mountains to the foothill wash at the base of the San Gabriel Mountains. Several waterfalls also exist above Eaton Falls, which are more secluded. This area is very popular with the public, especially during the summer months, for hiking and swimming.

CRAM scores averaged 70 at Eaton Canyon from 2009 to 2012. The buffer zones on this reach are relatively undisturbed and continuous and the hydrology upstream and downstream of the site is natural. The low structural diversity of the plants, as well as a large number of the invasive plant species contributed to lower scores.

HAINES CREEK POOLS AND STREAM (TUJUNGA PONDS WILDLIFE SANCTUARY; LALT407)

The 13-acre Tujunga Ponds in Sunland is a Caltrans mitigation project constructed following completion of the 210-Foothill Freeway. The site was acquired by Los Angeles County Parks and Recreation Department in 1978 and contains two small lakes and surrounding dense willow riparian and cottonwood riparian woodlands. The natural areas and existing trails around the ponds are used by visiting groups for nature study, photography, and similar passive recreation under permit from LACDPW.

CRAM scores averaged 69 from 2009 to 2012. The buffer zones surrounding the pools are in excellent condition, while their physical structure is not ideal owing to a lack of complexity. The hydrology of the pools is highly modified, albeit functional considering this is a remediated site. The biotic structure of the site is relatively good with several endemic species and good zonation.

ARROYO SECO (LALT450)

The Arroyo Seco site is located downstream of Devil's Gate Dam. The Arroyo Chub, a locally-extirpated native fish, was recently reintroduced to this section of the Arroyo Seco following habitat restoration. The CRAM score at this site averaged 65 between 2009 and 2010. This reach has a relatively good riparian zone and moderate stream bed complexity; however, surrounding residential development and heavy recreational use has adversely impacted the site.

3.5   POST-FIRE TARGET SITES

The 2009 Station Fire burned 160,577 acres of the Angeles National Forest including 68% of the upper watershed of the Los Angeles River. Three previously sampled random sites, reaches of Lynx Gulch, Big Tujunga, and Gould Mesa, were within the burn area and provided a unique opportunity for the LARWMP to monitor the post-fire decline and recovery of benthic macroinvertebrates and habitat. This adaptive monitoring strategy provided a timely response to the impacts of unanticipated (natural) disasters.

Gould Mesa, located on the Arroyo Seco, exhibited the greatest decline in biological condition following the fire (Figure 18). Prior to the burn, this site had one of the higher biological condition scores in the watershed (SoCA IBI = 79). Directly following the fire, the score was below the impairment threshold and has only slightly improved in three years. Biological condition at Big Tujunga has actually improved following the fire relative to pre-fire condition while Lynx Creek has not improved.

FIGURE 18. IBI scores at post-fire target sites.

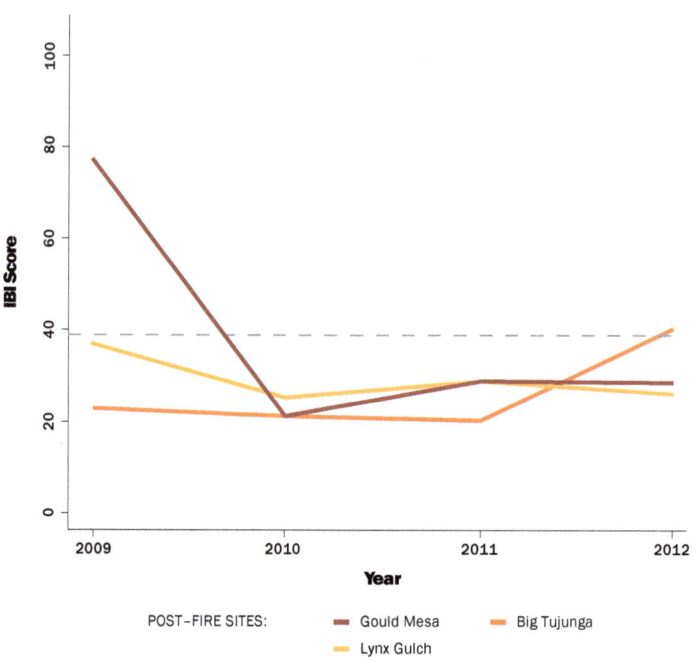

As might be expected, the riparian zones at all three of these sites were altered during the fire, as demonstrated by decreases in the riparian health scores (CRAM) at both Gould Mesa and Big Tujunga Wash (Figure 19). Loss of a healthy riparian zone after a fire leads to bank erosion and loss of canopy cover that are important for healthy BMI communities. Ongoing monitoring at these sites should yield important information regarding the recovery of post-fire stream reaches.

3.6   SUMMARY AND NEXT STEPS

Temporal trends in water chemistry could not be discerned at the confluence sites during the initial four years of monitoring; more years of monitoring will help to determine trends. The poor biological community condition at confluences reflects the nature of riparian habitat that has been highly modified in the lower watershed. Future research will explore the source of the recurrent chronic (reproductive) toxicity at the confluence of the Arroyo Seco with the Los Angeles River. Sediment quality in the estuary was highly variable across the four-year period, ranging from 'clearly impacted' to 'unimpacted'.

As expected, the physical and riparian habitats are healthier at sites in the upper watershed in the Angeles National Forest compared to lower watershed sites. To reduce the redundancy of conducting CRAM assessments annually, and acknowledging the slow rate of change in this indicator, sampling has been reduced to every three or four years.

Three previously sampled random sites were within the burn area of the 2009 Station Fire, providing a unique opportunity to monitor the post-fire decline and recovery of benthic macroinvertebrate communities and riparian habitat. The condition of the biological communities declined after the fire and has not recovered to pre-fire condition, as exemplified at Gould Mesa on the Arroyo Seco. Some of the reduction in biological condition can be attributed to loss of riparian habitat that is recovering at all sites. Continued monitoring at these locations provides important information regarding post-fire recovery of benthic macroinvertebrate communities.

Photo (left): Upper Arroyo Seco after the Station Fire in 2009.

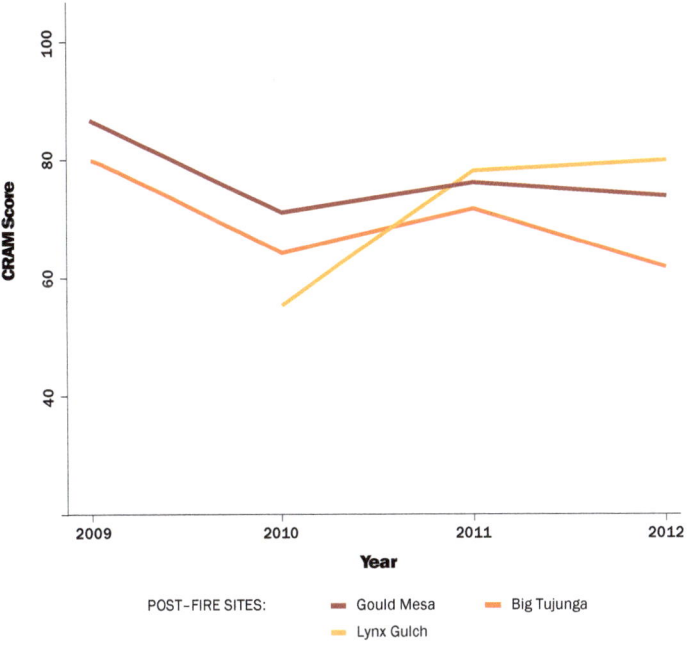

FIGURE 19. CRAM scores at post-fire target sites.

POST-FIRE SITES: ▬ Gould Mesa   ▬ Big Tujunga   ▬ Lynx Gulch

REFERENCES

SCCWRP, 2009. Sediment quality assessment DRAFT technical support manual. Southern California Coastal Water Research Project, Technical Report 582, May 2009. http://www.sccwrp.org/ResearchAreas/Contaminants/SedimentQualityAssessment.aspx.

Stein, E.D., Fetscher, A.E., Clark, R.P., Wiskind, A., Grenier, J.L., Sutula, M., Collins, J.N., and Grosso, C., 2009. Validation of a Wetland Rapid Assessment Method: Use of EPA's Level 1-2-3 Framework for Method Testing and Refinement. Wetlands 29(2): 648–665. http://www.bioone.org/doi/abs/10.1672/07-239.1.

FIGURE 20. Location of water reclamation plants in the Los Angeles River Watershed.

MAP DESIGNED BY THE COUNCIL FOR WATERSHED HEALTH, 2012

CHAPTER 4

ARE RECEIVING WATERS NEAR DISCHARGES MEETING WATER QUALITY OBJECTIVES?

Reclaimed water from Tillman plant discharging to Lake Balboa (Photo courtesy of John Schramm).

4.1 BACKGROUND

The type and magnitude of pollutants that enter the Los Angeles River is determined by discharge sources and season. During the dry season, the Los Angeles River is primarily sustained by treated water from water reclamation plants (point sources), and to a lesser degree, urban runoff and groundwater seepage (nonpoint sources). In contrast, stormwater runoff (nonpoint source) contributes the largest volumes during the rainy season.

The goal of this question is to assess the impact of known point source discharges on receiving water quality in the Los Angeles River (Figure 20). The first five years of the LARWMP focused on effluents from three publicly-owned treatment works (POTWs) that discharge tertiary-treated effluents

WATER QUALITY OBJECTIVES FOR RECEIVING WATERS

Nutrients, metals, *E. coli,* and organic constituents were compared to the objectives described in the Los Angeles Region Basin Plan for the Coastal Watersheds of Los Angeles and Ventura Counties or their individual State of California Toxics Rule (CTR) to determine if they were above either the acute or chronic thresholds. For some of these constituents, objectives are adjusted according to other measured parameters such as hardness for metals and pH and temperature for ammonia. Acute thresholds represent maximum 1-hr concentrations protective of aquatic life uses and the chronic thresholds represent maximum 30-day average concentrations protective of aquatic life uses.

Los Angeles Region Basin Plan for the Coastal Watersheds of Los Angeles and Ventura Counties
http://www.waterboards.ca.gov/ losangeles/water_issues/programs/ basin_plan/

AB 411
http://www.swrcb.ca.gov/water_ issues/programs/beaches/beach_ surveys/bills/ab_411_bill_19971008_ chaptered.pdf

CA CTR
http://www.epa.gov/region9/water/ ctr/

LOS ANGELES RIVER WATERSHED WATER RECLAMATION PLANTS

BURBANK WRP

The City of Burbank Water Reclamation Plant (BWRP) was built in 1966 and currently treats 9 million gallons of sewage per day (MGD). Before the BWRP was built, the City of Burbank sent all of its wastewater to the City of Los Angeles for treatment and disposal.

The plant was upgraded in 2000 to ensure that it could meet the new stringent treatment regulations mandated by the Los Angeles Regional Water Quality Control Board (LARWQCB). The plant was upgraded again in 2002 to convert ammonia to nitrate, a less toxic form of nitrogen.

Burbank Water Reclamation Plant. Photo courtesy of City of Burbank Public Works Department.

Los Angeles-Glendale Water Reclamation Plant. Photo courtesy of City of Los Angeles Bureau of Sanitation.

LOS ANGELES-GLENDALE WRP

Los Angeles-Glendale Water Reclamation Plant Water Reclamation Plant commenced operations in 1976 as the first water reclamation plant in the city. The cities of Los Angeles and Glendale co-own the plant, and the City of Los Angeles' Bureau of Sanitation operates and maintains it. Each city pays 50% of the costs and receives an equal share of the recycled water. The plant processes around 20 MGD of wastewater per day.

DONALD C. TILLMAN WRP

The Donald C. Tillman Water Reclamation Plant began continuous operation in 1985. A major construction project that doubled the capacity of DCT was completed in 1991 – expanding the plant from 40 MGD to 80 MGD.

The Tillman Plant, together with the Los Angeles-Glendale Water Reclamation Plant, is the leading producer of reclaimed water in the San Fernando Valley. By reclaiming a significant portion of wastewater, these treatment facilities have provided critical hydraulic relief to the City's downstream sewage lines.

Donald C. Tillman Water Reclamation Plant. Photo courtesy of John Schramm.

to the Los Angeles River above the confluence with the Arroyo Seco:

- Burbank Water Reclamation Plant (BWRP),

- Los Angeles-Glendale Water Reclamation Plant (LAGWRP)

- D.C. Tillman Water Reclamation Plant (DCTWRP)

The treatment capacities of these WRPs range from 9 million gallons per day (MGD) for the Burbank WRP to 20 MGD and 80 MGD for the Glendale and Tillman WRPs, respectively. Las Virgenes Municipal Water District (LVMWDs) Tapia Plant is also permitted to discharge 2 MGD to the Los Angeles River at certain times of year; however, discharges are typically much less than this *(see Table 1, page 2)*.

National Pollution Discharge Elimination System (NPDES) permits require these POTWs to monitor water quality upstream and downstream of the point of discharge to demonstrate that they attain certain water quality standards. LARWMP consolidated these data from 2008 to 2012 and compared them to the State of California threshold values considered to be protective of aquatic life.

LARWQCB also issues industrial NPDES permits and general industrial stormwater permits to facilities that discharge directly to the Los Angeles River and its tributaries. The largest numbers of dischargers occur in the cities of Los Angeles (many within the community of Sun Valley), Vernon, South Gate, Long Beach, Compton, and Commerce. Metal plating, transit, trucking &

warehousing, and wholesale trade are a large component of these businesses. Determining the impacts of these discharges on the River is a future goal of the LARWMP.

4.2 WATER QUALITY OF RECEIVING WATERS (2008-2012)

The results from five years of monitoring are summarized below and a more comprehensive analysis is provided in the LARWMP Annual Reports[1]. Effluent from the POTWs does cause a substantial change in the water quality of the Los Angeles River, for the better. Concentrations of dissolved metals, fecal indicator bacteria, and suspended solids in the effluents are often lower than concentrations in the Los Angeles River due to permit requirements. Thus, river concentrations of these constituents are consistently reduced downstream from the WRPs over the monitoring period *(Figure 21)*.

Despite the large number of exceedances both upstream and downstream of the Burbank POTW, *E.coli* concentrations were up to 98% lower below the discharge, as they were diluted by disinfected effluents. This trend was observed at both the Glendale and Tillman POTWs.

In contrast, the effluents from these facilities contain higher concentrations of nutrients (e.g., ammonia and nitrate) and disinfection by-products than the Los Angeles

---

1   http://watershedhealth.org/programsandprojects/larwmp.aspx

FIGURE 21. *E.coli* exceedances upstream and downstream of POTW effluents.

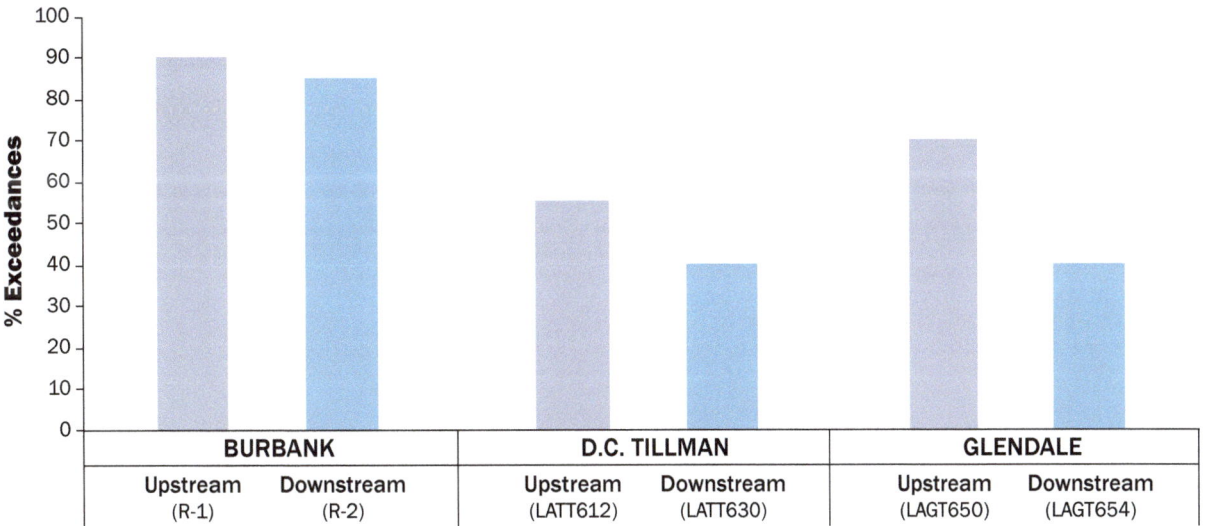

TABLE 13. WRP receiving water concentrations.

	Range (n)			Threshold Value
	Donald C. Tillman WRP	Glendale WRP	Burbank WRP	
Nitrate-Nitrogen (mg/L)	3.39-5.11 (12)	2.52-5.45 (12)	2.45-6.33 (12)	8
Ammonia-Nitrogen (mg/L)	ND-0.42 (12)	ND-0.44 (12)	0.11-1.97 (12)	*
THMs (ug/L)	2.48-3.38 (2)	2.21-3.07 (2)	3.2-22.8 (12)	180

ND: Not-detected
 *: calculated based on temperature and pH

River, so that these nutrients increase downstream from the WRPs *(Table 13)*. Despite this, nearly every constituent that was assessed against a water quality standard during the five-year period was within the thresholds regarded as safe for human or aquatic life. In instances when water quality standards were exceeded, they almost exclusively occurred above the WRP discharge locations.

The Cities of Burbank and Los Angeles and the LARWMP will continue to monitor receiving waters to determine if they are meeting the water quality objectives for their beneficial uses. This monitoring will determine changes in the concentration and presence of constituents as new chemicals are introduced, advance treatment processes are implemented, and policies restricting the use of substances are enacted. As mentioned above, LARWMP endeavors to evaluate the impact of industrial discharges on receiving water quality in the Los Angeles River in the future.

CHAPTER 5

IS IT SAFE TO SWIM?

5.1 BACKGROUND

When the public imagines the Los Angeles River and its watershed, they commonly visualize the concrete-lined channels of the lower watershed. This image overshadows the abundant recreational opportunities provided by the freshwater lakes and streams, particularly headwater streams in the Angeles National Forest. During the warm spring and summer months, thousands of locals and visitors enjoy swimming in cool waters of these relatively natural streams. Despite this popularity, prior to LARWMP, little was known about the levels of pathogenic bacteria at popular swimming sites throughout the watershed.

To determine the human health safety of swimming in these waters, LARWMP measured Fecal Indicator Bacteria (FIB), which includes *E. coli* and fecal coliform bacteria. By themselves these groups of bacteria typically do not cause illness; however, the presence of *E. coli* in recreational waters indicates fecal contamination by humans or animals and acts as a freshwater diagnostic tool for the presence of other, more harmful, pathogens such as *Salmonella* and *Giardia*. Although nearly all strains of *E. coli* are harmless to humans, enterohemorrhagic strains such as O157:H7 can cause bloody diarrhea, stomach cramps, nausea and vomiting and in more severe cases anemia, kidney failure and death in the elderly, the very young or the immunocompromised (Nataro, 1998; Keene, 1994).

In California, the State Water Board and Regional Water Quality Control Boards determine waters that are suitable for swimming (REC-1) and describe Water Quality Objectives (WQOs) to protect these waters (AB 411). In particular, at locations where people are in direct contact with the water, such as swimming and wading, bacterial pathogens should not exceed levels that pose a direct risk to human health. The Los Angeles Basin Plan *E. coli* standard for waters designated for recreational swimming is 235 per 100 mL based on a Most Probable Number (MPN) per single sample analysis (LARWQCB 1994). This standard was developed in line with human health guidelines that allow "historically acceptable illness rates," which for freshwater bodies has been designated as eight illnesses per 1,000 swimmers (US EPA, 1986).

Photo: (right) Hermit Falls swimming site in the Angeles National Forest.

5.2 SAMPLING AND SITE SELECTION

Weekly sampling for *E. coli* began in 2009 during the summer (May to September) at high-use recreational swimming sites. Sampling for indicator bacteria at non-body contact (REC2) locations at the confluences of major tributaries to the Los Angeles River and at the estuary was also initiated to determine the concentrations of indicator bacteria emanating from the watershed as a whole. The results of the latter sampling efforts are summarized in *Chapter 2 - What are the conditions of Streams in the Watershed?*

Twelve recreational swimming sites were monitored during 2009 thru 2012 *(Table 14),* including streams in the Angeles National Forest and lakes and streams in the lower watershed. Initially, sites were selected based on the collective knowledge of the workgroup of popular swimming locations. Sites were added or excluded as the LARWMP improved its understanding of the recreational use of these lakes and streams. Specifically, in 2012, two sites in the Rio Hondo were excluded due to lack of observed recreational use during the previous 3-years of monitoring, and Hermits Falls was added in the Angeles National Forest. In addition, sampling was suspended in 2010 at sites in the upper watershed that were closed following the 2009 Station Fire. These sites were reopened in 2011 where sampling continues.

To elucidate the relationships between heavy recreational use and *E. coli* concentrations, sampling was concentrated around weekends and holidays when the swimming intensity is greatest. Depending

on the site, sources of indicator bacteria and pathogen contamination could include humans, dogs, wildlife, urban runoff, and refuse from campgrounds.

5.3 RESULTS

During the summer months of 2009 thru 2012, a total of 535 samples were collected from swimming sites and analyzed for *E. coli (Table 15).* 21% of these samples exceeded the REC-1 standard for a single sample for *E. coli* (235 MPN/100 mL) *(Figure 22).* This is higher than in the neighboring San Gabriel River watershed where 4% of samples exceeded the REC-1 standard during 2007 through 2009.

The greatest frequency of REC-1 exceedances occurred at the two sites adjacent the Rio Hondo Bike path in the Whittier Narrows recreation. The Upper Rio Hondo Site (LALT201) is a small lake behind the Whittier Narrows Golf Course that receives Alhambra wash discharge and

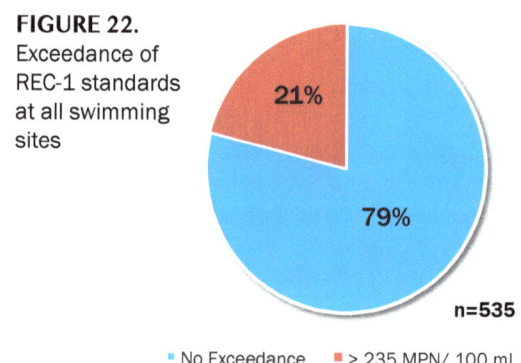

FIGURE 22. Exceedance of REC-1 standards at all swimming sites

n=535

■ No Exceedance   ■ > 235 MPN/ 100 mL

TABLE 14. Recreational swimming sites in the LARWMP

SWIMMING SITES	SITE CODES	NOTES
Bull Creek, Sepulveda Basin	LALT200	
Eaton Canyon	LALT204	
Gould Mesa Campground	LAUT209	2010 Station Fire closure-reopened 2011
Hermit Falls	LAUT213	Commenced sampling in 2011
Millard Campground	LAUT203	Waterfall closed 2010
Peck Rd Water Conservation Park	LALT212	
Sturtevant Falls	LAUT210	
Switzer Falls Day-Use Area	LAUT208	2010 Station Fire closure-reopened 2011
INACTIVE SITES		
Big Tujunga Delta Flat	LAUT206	2010 Station Fire closure
Hidden Springs (upper Tujunga)	LAUT211	2010 Station Fire closure
Upper Rio Hondo	LALT201	Sampling ceased in 2011 due to no observable human use
Bosque del Rio Hondo	LALT205	Sampling ceased in 2011 due to no observable human use

TABLE 15. Exceedances of REC-1 water standards for *E. coli* at swim sites from May to September, 2009 thru 2012. REC-1 Standards: 30-day geometric mean = 126 MPN/100 mL; single sample = 235 MPN/100 mL. Red indicates exceedance of REC-1 standard.

SWIM SITES	SAMPLING YEAR	GEOMETRIC MEAN						SINGLE SAMPLE EXCEEDANCES	
		May	Jun	Jul	Aug	Sep	n =	#	%
Big Tujunga Delta Flat	2009		57	29	10		13	0	0%
Bull Creek, Sepulveda Basin	2009		109	185	139	86	8	2	25%
	2010		61	321	85	257	15	5	33%
	2011	110	346	112	86	699	15	4	27%
	2012	154	174	448	165	97	20	7	35%
Eaton Canyon	2009		136	32	27		13	2	15%
	2010	2420	125	129	86	2758	17	5	29%
	2011	86	133	195	64	10005	15	3	20%
	2012	680	49	243	363	1485	20	7	35%
Gould Mesa Campground	2009		32	79	54		13	0	0%
	2011	20	39	31	20	31	16	0	0%
	2012	10	27	42	33	13	20	0	0%
Hidden Springs (upper Tujunga)	2009		24	10	10		13	0	0%
Millard Campground	2009		101	44	62		13	0	0%
	2010		46	104	63	61	15	1	7%
	2011	31	64	20	42	26	16	0	0%
	2012	43	31	14	31	10	20	0	0%
Peck Rd Water Conservation Park	2009			40	43	10	10	1	10%
	2010	41	121	141	38	44	17	4	24%
	2011		1579	357	142	488	13	7	54%
	2012	68	636	329	204	63	20	5	25%
Sturtevant Falls	2009		31	11	47		13	1	8%
	2010	10	118	25	31	284	17	1	6%
	2011	15	34	717	148	140	16	4	25%
	2012	111	78	202	108	80	20	4	20%
Hermit Falls	2011	25	20	80	45	115	17	0	0%
	2012	10	283	271	62	166	20	5	25%
Switzer Falls Day-Use Area	2009		57	270	194		13	5	38%
	2011			52	56	128	8	0	0%
	2012	10	264	121	153	231	20	4	20%
Upper Rio Hondo	2009		706	1571	420	909	14	11	79%
	2010	31	29	148	56	3169	17	4	24%
	2011	503	1870	330	280	20750	16	8	50%
Bosque del Rio Hondo	2009		134				3	2	67%
	2010		76				3	0	0%
	2011	86	6247	163	631	24200	16	8	50%
						TOTALS	535	110	21%

LOS ANGELES RIVER RECREATIONAL ZONES

Recently, soft-bottomed portions of the Los Angeles River in the lower watershed have been recognized as valuable recreational assets and pilot studies are underway to determine their suitability as recreational zones.

In 2010, Los Angeles Conservation Corps (LACC) along with its partners formed the Paddle the LA River pilot program. This is the first non-motorized boating program on the Los Angeles River at Sepulveda Recreation Basin. The program partners expect that by paddling this scenic stretch, people experience first-hand that Los Angeles River is part of an ecosystem that is both beautiful and significant to Los Angeles' past and future. This program operated for two years, but did not receive an Army Corps permit for 2013.

In 2013, The Mountains Recreation and Conservation Authority (MRCA), in cooperation with the City and County of Los Angeles, the Los Angeles County Flood Control District, and the Army Corps of Engineers, administered the Los Angeles River Recreation pilot program to increase safe public access to the L.A. River and to promote the goal of river revitalization.

The Los Angeles River Recreation Zone provides an opportunity for any member of the public to walk, fish, and kayak on a 2.5-mile portion of the L.A. River in Elysian Valley from Memorial Day (May 27, 2013) to Labor Day (September 2, 2013) and from sunrise to sunset daily, during safe conditions. Both programs received permits for the 2014 season, thus providing opportunities for people to experience these two stretches of the LA River.

Bosque Del Rio Hondo (LALT205) is located downstream near Bosque Del Rio Hondo Park. Nearly half of the samples collected at these two locations exceeded REC-1 standards. Body contact recreation was observed at the Upper Rio Hondo Site prior to the commencement of the monitoring program; however, no humans were observed at these sites during LARWMP sampling. For this reason, sampling was discontinued at these sites in 2012.

At the most popular swimming sites in the Angeles National Forest (Figure 23), the frequency of single-sample REC-1 exceedances ranged between 14% and 26%, at Hermit Falls, Eaton Canyon, Switzer Falls, and Sturtevant Falls. Elevated E. coli levels were frequently measured at Bull Creek and Peck Rd Water Conservation Park; however, body contact by humans was observed only occasionally and sporadically at these sites throughout the monitoring period. Moreover, sampling was conducted between 9 am and 2 pm when recreational use was expected to be highest; therefore, LARWMP did not record recreational use occurring outside of this period.

The 30-day geometric mean provides an indication of the persistence of elevated bacterial levels at a site. The State of California REC-1 bathing water standards specify that at least five samples should be collected per month per site before the 30-day geometric mean standard can be applied. The 30-day geometric average standard (126 MPN/100 mL) was frequently exceeded at most sites during the monitoring period. The exceptions were sites in the Angeles National Forest on Big Tujunga Canyon Rd and Angeles Crest Highway, and include sites closed following the 2009 Station Fire.

At most sites other than those in the Angeles National Forest, the frequency of exceedances increased in 2011-2012; this was accompanied by an observable decrease in water level compared to previous years. Recreational use is heavier towards the end of most summers. Higher bacteria levels were common on weekends and holidays when hundreds of people can be observed swimming and wading in these streams (Table 15).

5.3.1 SOURCES OF E.COLI

Swim sites in the Angeles National Forest are very popular with the public during the warm summer months. These sites are readily accessible to the nearly 10 million[1] residents of Los Angeles County and provide an opportunity to experience relatively undisturbed streams

---

1 http://quickfacts.census.gov/qfd/states/06/06037.html. Retrieved 08/15/2013

FIGURE 23. Exceedance of REC-1 standards at swimming sites in the upper watershed

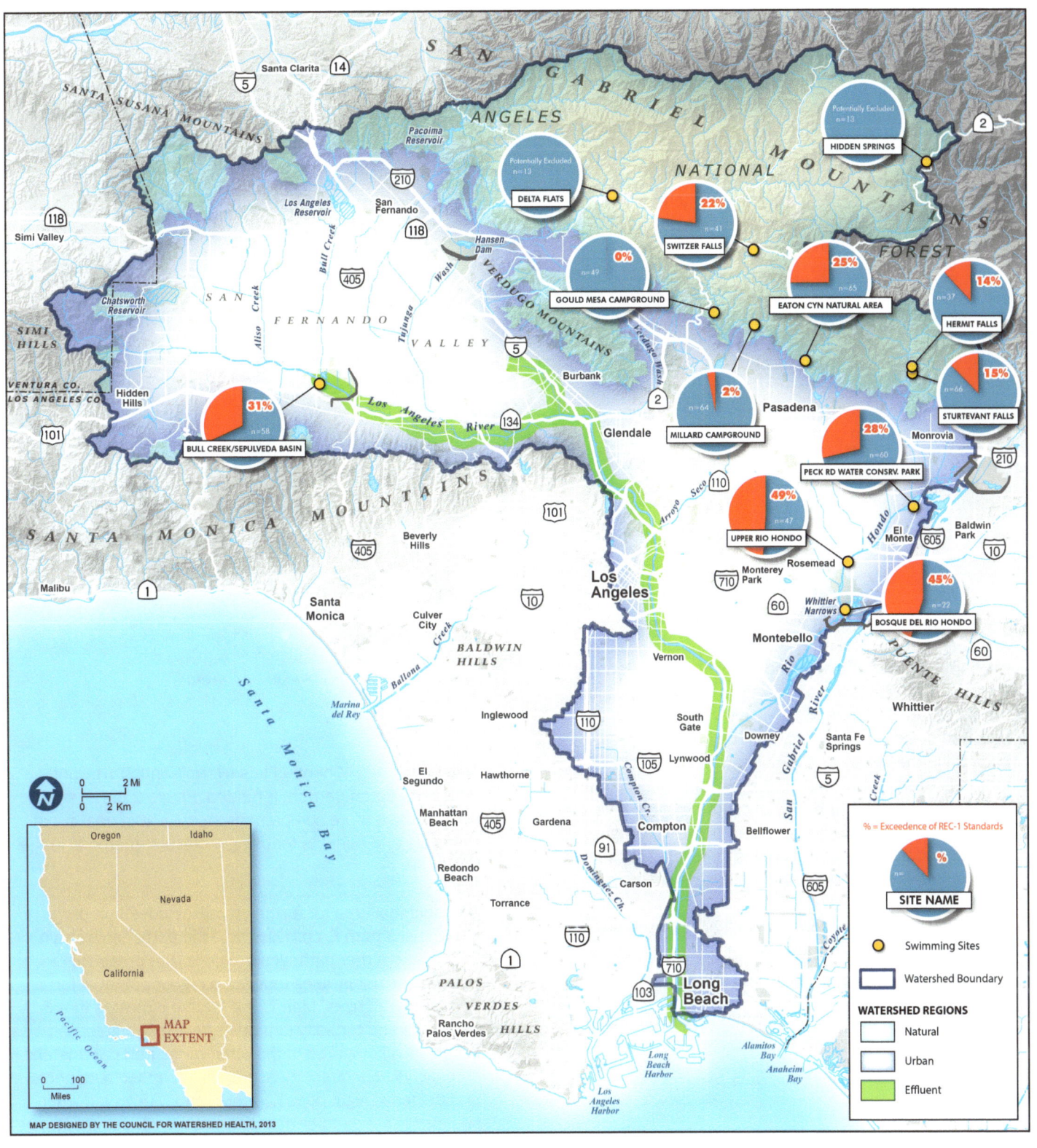

in an otherwise highly urbanized watershed. In 2012, Hermit Falls, Eaton Canyon, and Sturtevant Falls were the most popular swimming sites.

In an attempt to discern possible sources of *E. coli* at swimming sites, LARWMP records numbers of humans, dogs, and birds at each site during sample collection. Although these counts represent only a snap-shot of recreational activity during a sampling event, eventually they can suggest relationships between *E. coli* concentrations and use. Turbidity, water temperature, air temperature, pH, and conductivity are also recorded during sample collection.

We compared *E. coli* counts to these above-mentioned parameters. The strongest correlation was between turbidity and *E. coli* (r = 0.32). This may indicate that these bacteria may reside in bottom sediments and are dispersed into the water column by swimming activity. These results also show that there is a large amount of variation inherent in bacteria data sets, which makes tracking elevated concentrations to their sources difficult. Future monitoring will address the sources of fecal indicator bacteria and this is described below.

5.4   SUMMARY AND NEXT STEPS

Swimming and wading at freshwater sites in the Los Angeles River Watershed is popular, particularly during the summer months and on holiday weekends. People leave behind trash, including likely fecal contamination from dogs and toddlers. Wading and swimming stir up sediments and disturb the stream banks that increases bacteria in the water. The sampling and analytical methods used, however, do not allow for confirmation of potential sources for bacterial contamination. As a result, additional study is required before making recommendations as to best management practices to reduce bacterial contamination at these swim sites.

In 2012, Council for Watershed Health, in partnership with staff from the Jet Propulsion Laboratory and Duke University, was awarded a Robert & Patricia Switzer Foundation grant to conduct a pilot study to determine the feasibility of using molecular methods in the Los Angeles Watershed for risk assessment and source identification. SCCWRP was subsequently included as a project partner.

Phase 1 of the study identified a number of deficiencies in our knowledge of recreational use of streams in the Angeles National Forest. In response, LARWMP will conduct recreational-use surveys at two of the most popular fresh water swimming sites in the Angeles

Sturtevant Falls swimming site in the Angeles National Forest.

National Forest—Chantry Flats (Hermit and Sturtevant Falls) and Eaton Canyon. If funded, Phase 2 of the study will include a more comprehensive economic analysis of the costs and benefits for using molecular methods in the LARWMP.

An additional area for further study includes human health risks from *E. coli* O157:H7, the pathogenic strain of *E. coli*, or other pathogenic strains. The presence of *E. coli* does not by itself indicate a higher potential for human health risks. Monitoring to-date indicates that *E. coli* levels and turbidity demonstrate the strongest co-occurrence compared to the other measured parameters. Turbidity is an indirect measure of suspended sediments and bacteria might become more readily detected when sediments are disturbed and resuspended. Sediments can serve as a reservoir and growth media for bacteria, including pathogenic strains. A favorable environment for bacteria is created by the availability of soluble organic matter and nutrients, protection from predators such as protozoa, and shielding from exposure to the UV sunlight

(Kim, et al., 2010). These relationships will be further investigated in future monitoring and research.

The LARWMP continues to revise the Safe to Swim program to ensure that watershed managers can understand and prioritize watershed-scale issues in a timely and efficient manner. Sampling will start at the non-chlorinated natural Lake at Hansen Dam in 2013 where people are regularly observed swimming.

It is a priority of LARWMP to make monitoring data available through the LARWMP portal *(http://108.168.216.185:86/la-portal/)*. In 2013, LARWMP Safe to Swim data is also available through the Waterkeeper Alliance Swim Guide App. LARWMP is also working to have its data included in the SWRCB Safe to Swim portal in the near future.

LARWMP Safe to Swim monitoring data is available through the Waterkeeper Alliance Swim Guide Mobile Application — theswimguide.org

REFERENCES

AB 411. California Assembly Bill No. 411. http://www.swrcb.ca.gov/water_issues/programs/beaches/beach_surveys/bills/ab_411_bill_19971008_chaptered.pdf

Keene, W.E., McAnulty, J.M., Hoesly, F.C., Williams, L.P., Hedberg, K., Oxman, G.L., Barrett, T.J., Pfaller, M.A., Fleming, D.W., 1994. A Swimming-Associated Outbreak of Hemorrhagic Colitis Caused by *Escherichia coli* O157:H7 and Shigella Sonnei, New England Journal of Medicine, 331:579-584.

Kim, J.W., Pachepsky, Y.A., Shelton, D.R., Coppock, C., 2010. Effect of streambed bacteria release on *E.coli* concentrations: Monitoring and modeling with the modified SWAT, Ecological Modelling 221 (2010) 1592–1604.

LARWQCB, 1994. Water Quality Control Plan, Los Angeles Region. Los Angeles Regional Water Quality Control Board, Los Angeles, CA.

Nataro J.P., Kaper J.B., January 1998. Diarrheagenic *Escherichia coli*. Clinical Microbiology Reviews 11(1): 142–201.

United States Environmental Protection Agency (USEPA), 1986. "Ambient Water Quality Criteria for Bacteria – 1986." EPA440/5-84-002.

FOLAR FISH STUDY

In the late summer and fall of 2007 the Friends of the Los Angeles River surveyed fish populations in the Glendale Narrows area, approximately an eight-mile stretch of natural bottom river that extends from Riverside Drive near Griffith Park to the Figueroa Bridge in Cypress Park.

Eight species of fish were collected: fathead minnow, carp, black bullhead, Amazon sailfin catfish, mosquitofish, green sunfish, largemouth bass, and tilapia. Mosquitofish and tilapia were the most abundant species.

The levels of mercury and PCBs of four composite samples of bullhead catfish, carp, sunfish, and tilapia were well below the three servings per week consumption guidelines described by OEHHA.

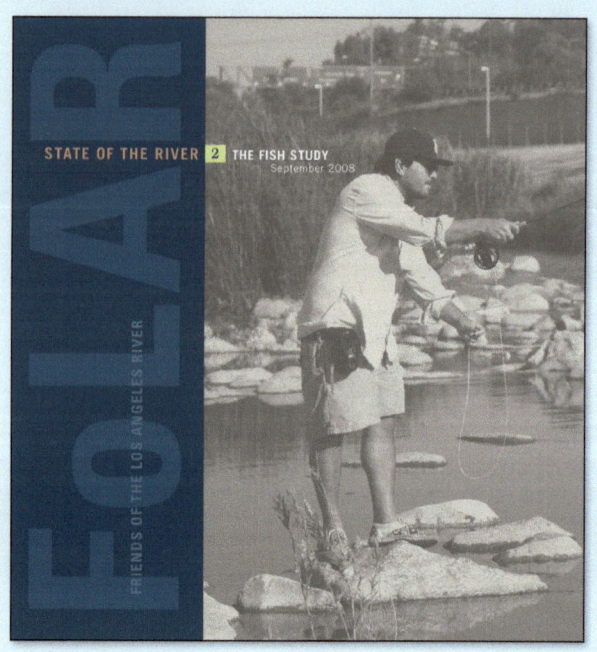

SURFACE WATER AMBIENT MONITORING PROGRAM (SWAMP) STUDIES

The SWAMP report, *Contaminants in Fish from California Lakes and Reservoirs, 2007-2008*, summarizes the results of a 2-year screening study of 272 of California's more than 9,000 lakes and reservoirs. This represents the beginning of a long-term, statewide, comprehensive bioaccumulation monitoring program for California surface waters.

The survey identified problems in certain areas of the state, with methylmercury and polychlorinated biphenyls (PCBs) being the contaminants of greatest concern. Methylmercury poses the most widespread potential health risk—21% of the lakes surveyed had at least one fish species with an average methylmercury level high enough (> 0.44 ppm) that OEHHA would consider recommending no consumption.*

Despite this, the degree of methylmercury contamination in the state's lakes is not unusual and is comparable to the average condition observed across the U.S. in a recent national lakes survey.

The study provides information that will be valuable in prioritizing lakes in need of further study to support development of consumption guidelines and cleanup plans.

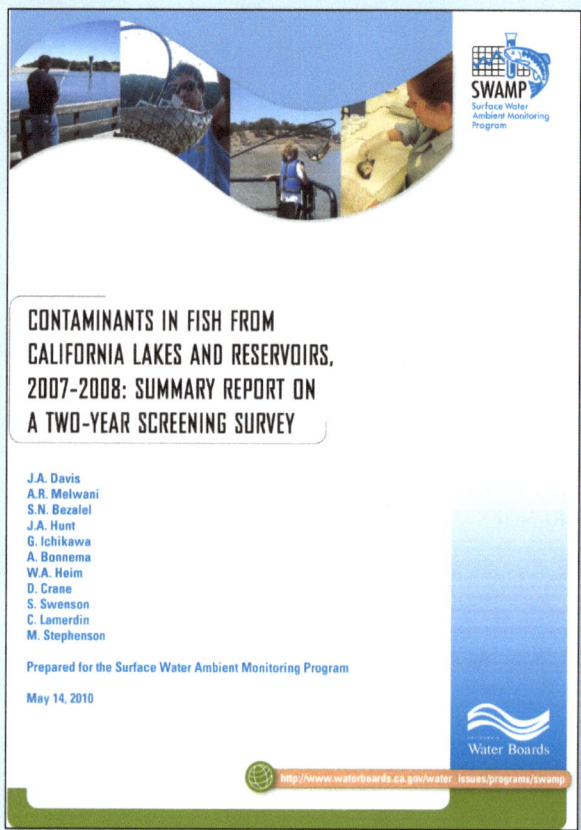

*For women between 18 and 45 years of age and children between 1 and 17 years of age

ARE FISH SAFE TO EAT?

FISH COLLECTED

2007 to 2012

LARWMP (2009, 2011-12): **96**

SWAMP (2007-10): **297**

FoLAR (2007): **1189**

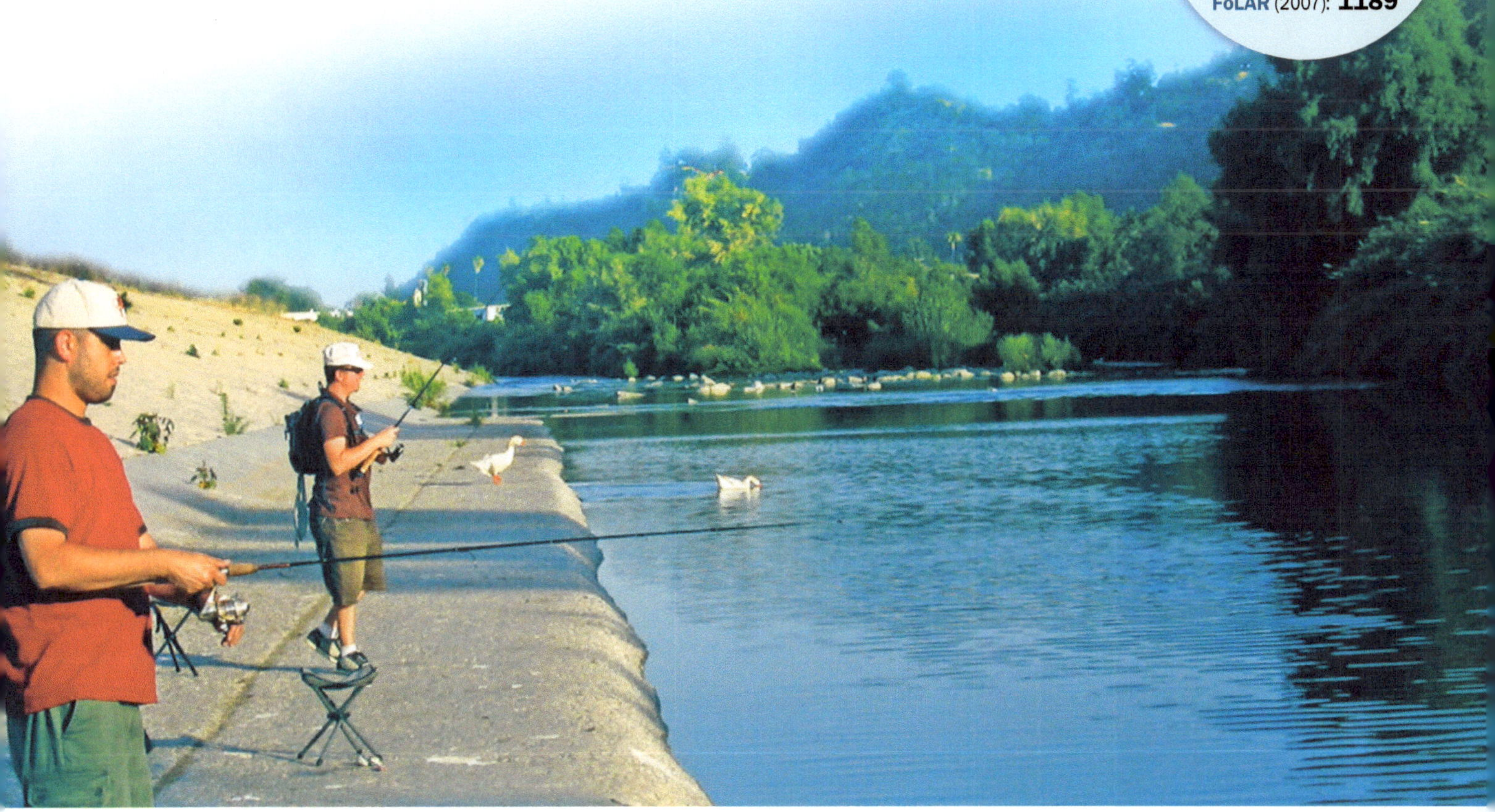

Fishing on the Los Angeles River along the Glendale Narrows.

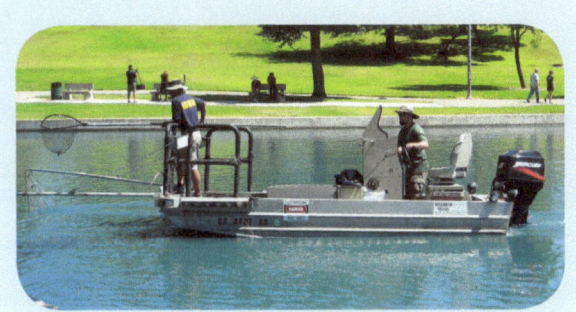

Fish tissues were collected following guidelines established by OEHHA (2005) using a combination of techniques depending on the water body and included boat drawn seines, hand seines, hook and line, and electro shocking.

http://oehha.ca.gov/fish/pdf/fishsampling121406.pdf

6.1 BACKGROUND

Prior to the start of the bioaccumulation sampling, little was known regarding the safety of eating fish caught in the watershed's estuary, creeks, and lakes. The Workgroup selected target species *(Figure 24)* and fishing locations *(Figure 25)* where fish were most likely being consumed based on a 2005 regional survey of anglers (Allen, *et al.*, 2008). Initially, largemouth bass and common carp were the most commonly caught fish. Blue gill, channel catfish, and redear sunfish were later targeted since they are more frequently eaten and little or no data were available for these fish.

Two fish tissue monitoring studies were conducted in summer 2007 by California's Surface Water Ambient Monitoring Program (SWAMP) and the Friends of the Los Angeles River (FOLAR). The LARWMP was designed to leverage and complement these studies, highlighting the goal of LARWMP to reduce redundancies in monitoring and to coordinate with existing programs.

FIGURE 24. Species of fish collected during 2006-2012 and average size

BLUE GILL (*Lepomis macrochirus*)

CHANNEL CATFISH (*Ictalurus punctatus*)

COMMON CARP (*Cyprinus carpio*)

LARGEMOUTH BASS (*Micropterus salmoides*)

REDEAR SUNFISH (*Lepomis microlophus*)

WHITE CROAKER (*Genyonemus lineatus*)

TILAPIA (*Oreochromis sp.*)

Illustrations:
Duane Raver, USFWS (bluegill, carp, bass, sunfish)
Diane Rome Peebles (striped mullet)
Joe Tomeilleri (catfish)
Jón Baldur Hlíðberg (tilapia)

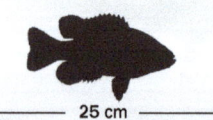

|← 25 cm →|

Fish are proportionately scaled by average size.

FIGURE 25. Fish tissue bioaccumulation sampling locations.

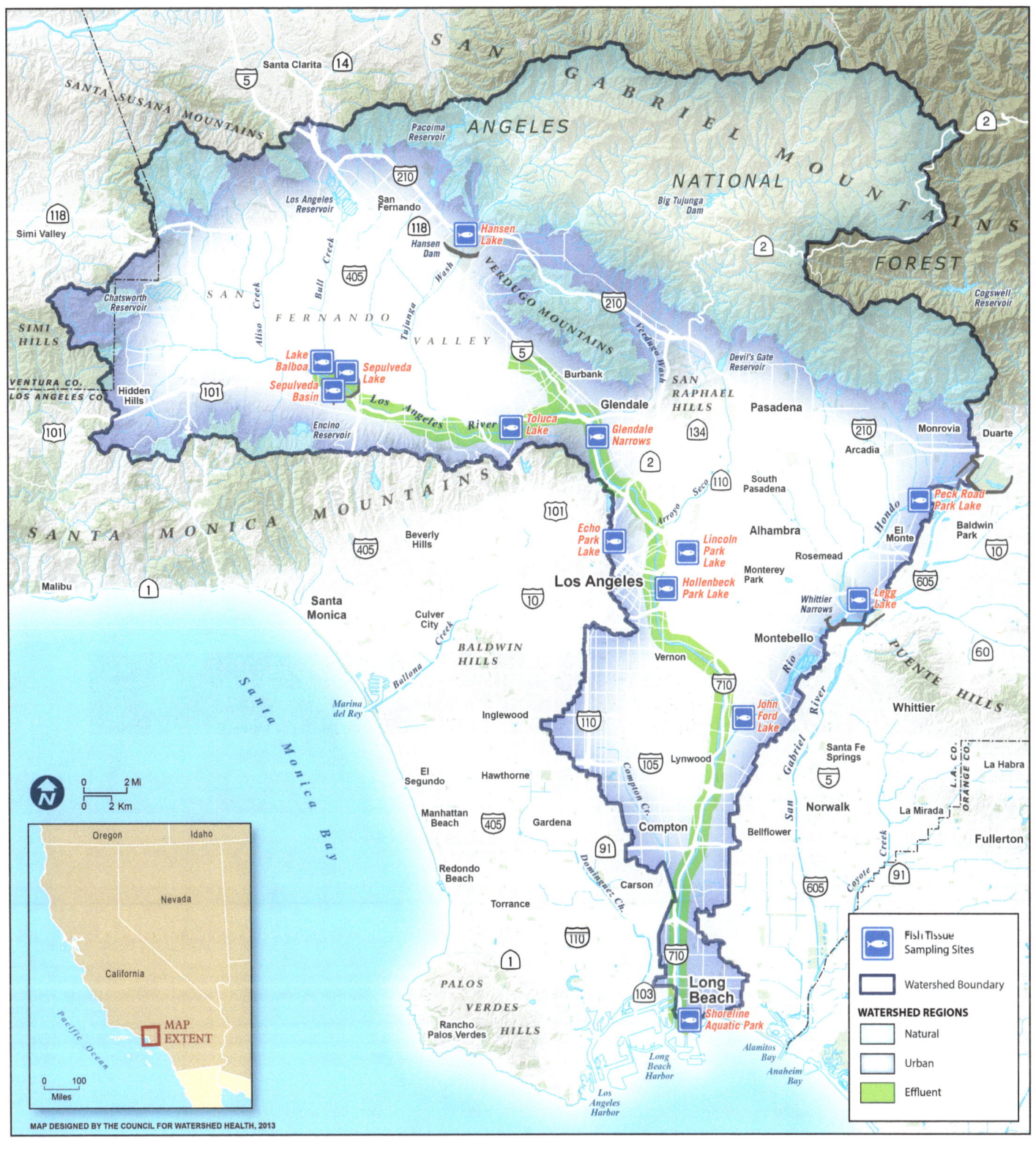

OEHHA ADVISORY TISSUE LEVELS

The OEHHA Advisory Tissue Levels (ATLs) were developed with the recognition that there are unique health benefits associated with fish consumption and that the advisory process should be expanded beyond a simple risk paradigm in order to best promote the overall health of the fish consumer. ATLs protect consumers from being exposed to more than the average daily reference dose for non-carcinogens or to a lifetime cancer risk level of 1 in 10,000 for fishermen who consume an 8-ounce fish fillet containing a given amount of a specific contaminant.

http://www.oehha.ca.gov/fish/so_cal/index.html

TABLE 16. OEHHA advisory tissue levels *(parts per billion)*

CONTAMINANT	THREE 8-OUNCE SERVINGS* A WEEK	TWO 8-OUNCE SERVINGS* A WEEK	ONE 8-OUNCE SERVINGS* A WEEK	NO CONSUMPTION
DDTs[1]**	≤520	>520-1,000	>1,000-2,100	>2,100
Methylmercury (Women aged 18-45 years and children aged 1-17 years)[1]	≤70	>70-150	>150-440	>440
Methylmercury (Women over 45 years and men)[1]	≤220	>220-440	>440-1,310	>1,310
PCBs[1]	≤21	>21-42	>42-120	>120
Selenium[2]	≤2500	>2500-4,900	>4,900-15,000	>15,000

[1]ATLs are based on non-cancer risk

[2]ATLs are based on cancer risk

*Serving sizes are based on an average 160-pound person. Individuals weighing less than 160 pounds should eat proportionately smaller amounts (for example, individuals weighing 80 pounds should eat one 4-ounce serving a week when the table recommends eating 8-ounces).

**ATLS for DDTs are based on non-cancer risk for two and three servings per week and cancer risk for one serving per week.

Four contaminants were selected for analysis based on their contribution to human health risk in California's coastal and estuarine fishes: mercury, selenium, total DDTs, and total PCBs. Fish tissue concentrations were evaluated using thresholds developed by the California Office of Environmental Health Hazard Assessment (OEHHA) for methylmercury, PCBs, DDTs, and selenium (Table 16).

6.2 CONCENTRATION OF CONTAMINANTS IN FISH TISSUES

Analysis of fish tissue indicated that, in general, fish that are commonly caught and consumed from lakes and streams in the watershed are likely to be safe to eat in moderate amounts. Three fish species (tilapia, redear sunfish, and bluegill) did not exceed OEHHA consumption thresholds (OEHHA 2008) for women of child-bearing age or children during the four-year period. Selenium and DDTs did not exceed consumption limitation thresholds for any fish species. There are important qualifications, however, for specific species caught in specific locations.

Mercury concentrations measured in largemouth bass and common carp captured from Hansen Lake, Legg Lakes, and Peck Road Water Conservation Park suggest these fish are in the "no consumption" category because they had concentrations above the threshold for one meal per week human consumption for women aged 18-45 years and children aged 1-17 years. Species more commonly consumed such as bluegill and catfish were well below all consumption thresholds (Figure 26).

Largemouth bass and common carp captured in Echo Park Lake, John Ford Lake, and Peck Road Water Conservation Park contained PCBs at concentrations suggesting that their consumption be limited to one or two meals per week. White croaker caught in the upper estuary contained PCB concentrations suggesting that their consumption be limited to one meal per week (Figure 27). Since 2009, OEHHA has issued updated fishing advisories for consumption of white croaker and other fish caught off Palos Verdes and other Southern California sites.

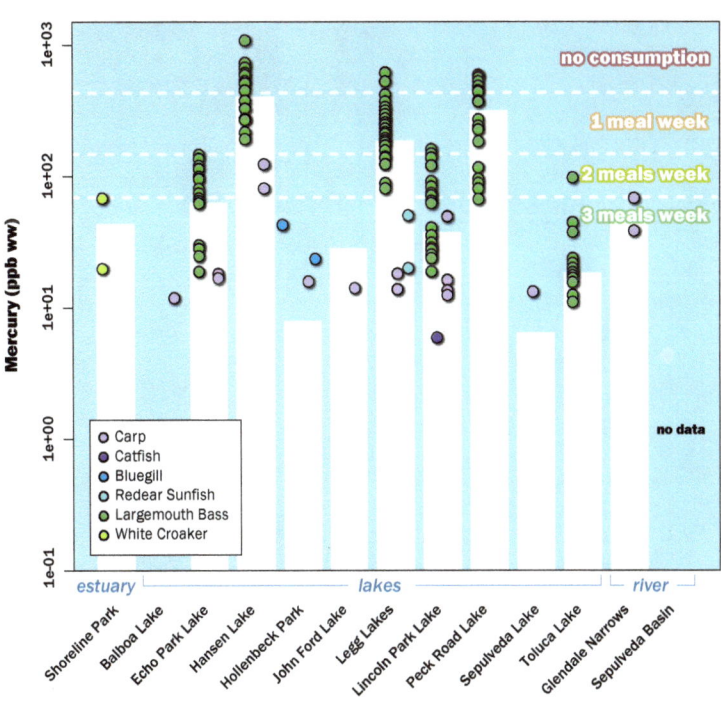

FIGURE 26. Total mercury in fish tissues.

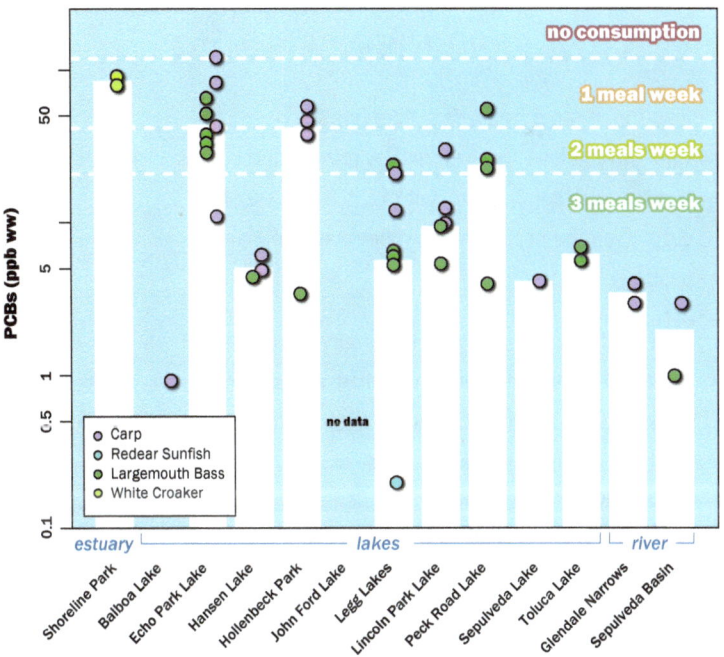

FIGURE 27. Total PCBs in fish tissues.

Photo (above): Fly fishing on the Los Angeles River near Atwater Village. (Courtesy of William Preston Bowling)

LOS ANGELES RIVER FISHING

Fishermen are a frequent sight along the banks of the Los Angeles River, where they use canned corn, tortillas, and commercial bait to catch fish, mostly carp. Even the occasional fly fisherman can be seen casting along the river.

Historically, the Los Angeles River supported a seasonal recreational fishery, with an annual winter run of steelhead trout. Following its channelization, the trout disappeared, replaced by carp, tilapia, and other non-native species.

Today, although fishing in the river is not an officially-sanctioned activity, and it is currently illegal to walk in the river channel below the bike paths, officials rarely cite the many anglers regularly seen along the soft-bottom sections where fish are to be found.

6.3 SUMMARY AND NEXT STEPS

In general, fish in the Los Angeles River Watershed had lower concentrations of mercury and comparable concentrations of selenium, DDT, and PCB when compared to fish from other parts of California (Davis, et al., 2010). Mercury concentrations in largemouth bass collected from streams nationwide, for example, (Scudder, et al., 2009) far exceeded those of bass measured in the Los Angeles River Watershed.

The LARWMP Workgroup decided after the fourth year to increase the number of composite samples collected annually from water bodies where elevated concentrations of mercury and PCBs were detected by program monitoring. Multiple fish species will also be collected from these water bodies to help assess which species are safe to eat.

To improve our understanding of angling and fish consumption behavior, we will continue conducting surveys of anglers at fishing locations throughout the watershed. These detailed angler surveys will provide valuable information to watershed managers on which species of fish are commonly caught and eaten, the frequency and quantity of consumption, and how they are prepared, as well as who is eating the fish.

We will work with OEHHA staff in the next phase of the program to develop a process for developing and posting fish advisories at lakes where there is a potential health risk for consuming contaminated fish.

REFERENCES

Allen, M.J, Jarvis, E.T., Raco-Rands, V., Lyon, G., Reyes, J.A., Petschauer D.M., 2008. Extent of fishing and fish consumption by fishers in Ventura and Los Angeles County watersheds in 2005. Technical Report 574. Southern California Coastal Water Research Project. Costa Mesa, CA.

Davis, J.A., A.R. Melwani, S.N. Bezalel, J.A. Hunt, G. Ichikawa, A. Bonnema, W.A. Heim, D. Crane, S. Swenson, C. Lamerdin, and M. Stephenson. 2010. Contaminants in Fish from California Lakes and Reservoirs, 2007-2008: Summary Report on a Two-Year Screening Survey. A Report of the Surface Water Ambient Monitoring Program (SWAMP). California State Water Resources Control Board, Sacramento, CA.

OEHHA (Office of Environmental Health Hazard Assessment), 2005. General protocol for sport fish sampling and analysis. Gassel, M. and R.K. Brodberg. Pesticide and Environmental Toxicology Branch, Office of Environmental Health Hazard Assessment, California Environmental Protection Agency. pp. 11.

OEHHA, 2008. Development of fish contaminant goals and advisory tissue levels for common contaminants in California sport fish: chlordane, DDTs, dieldrin, methylmercury, PCBs, selenium, and toxaphene. Pesticide and Environmental Toxicology Branch, Office of Environmental Health Hazard Assessment, California Environmental Protection Agency. pp. 115.

Scudder, B.C., Chasar, L.C., Wentz, D.A., Bauch, N.J., Brigham, M.E., Moran, P.W., and Krabbenhoft, D.P., 2009. Mercury in fish, bed sediment, and water from streams across the United States, 1998–2005: U.S. Geological Survey Scientific Investigations Report 2009-5109, pp. 74.

GOALS 2013 TO 2017

This State of the Watershed Report is the culmination of the first five years of a successful, cooperative watershed scale-monitoring program that seeks to address important management questions. Over the next five years, our primary goal is to characterize and understand the causes of impairment to biological communities including climate change. We will initiate studies to gain a better understanding of the sources of elevated bacteria levels at popular swimming locations, as well as characterizing the mercury concentrations in fish, and assist in developing any necessary fish advisories. Importantly, the program will continue to provide valuable ambient monitoring data to inform watershed-scale management actions.

PARTNERS

The following participants contributed staff time, laboratory analyses, and funding in a collaborative effort that included representatives from regulated, regulatory, environmental, and research organizations: Los Angeles, Burbank, Downey, Los Angeles Regional Water Quality Control Board, Los Angeles County Flood Control District, Southern California Coastal Water Research Project, and U.S. EPA. The cities of Los Angeles and Burbank provided a majority of the funding and the City of Los Angeles Environmental Monitoring Division provided extensive laboratory analyses.

Participating consultants included Aquatic Bioassay and Consulting Laboratories, Weston Solutions, IIRMES, Physis Environmental Laboratories Inc. Council for Watershed Health served as the program manager.